MOLECULAR EVOLUTION ON
RUGGED LANDSCAPES

MOLECULAR EVOLUTION ON RUGGED LANDSCAPES: PROTEINS, RNA AND THE IMMUNE SYSTEM

THE PROCEEDINGS OF THE WORKSHOP ON APPLIED MOLECULAR EVOLUTION AND THE MATURATION OF THE IMMUNE RESPONSE, HELD MARCH, 1989 IN SANTA FE, NEW MEXICO

Editors

Alan S. Perelson
Theoretical Division
Los Alamos National Laboratory

Stuart A. Kauffman
University of Pennsylvania and
the Santa Fe Institute

Proceedings Volume IX

SANTA FE INSTITUTE
STUDIES IN THE SCIENCES OF COMPLEXITY

CRC Press
Taylor & Francis Group
Boca Raton London New York

CRC Press is an imprint of the
Taylor & Francis Group, an **informa** business

Director of Publications, Santa Fe Institute: *Ronda K. Butler-Villa*
Technical Assistant, Santa Fe Institute: *Della Ulibarri*

First published 1991 by Westview Press

Published 2018 by CRC Press
Taylor & Francis Group
6000 Broken Sound Parkway NW, Suite 300
Boca Raton, FL 33487-2742

CRC Press is an imprint of the Taylor & Francis Group, an informa business

Visit the Taylor & Francis Web site at
http://www.taylorandfrancis.com

and the CRC Press Web site at
http://www.crcpress.com

Library of Congress Cataloging-in-Publication Data

Workshop on Applied Molecular Evolution and the Maturation of the
Immune Response (1989 : Santa Fe, N.M.)
Molecular evolution on rugged landscapes : proteins, RNA, and the immune
system : the proceedings of the Workshop on Applied Molecular Evolution and the Maturation of the Immune Response, held March, 1989 in
Santa Fe, New Mexico / editors, Alan S. Perelson, Stuart A. Kauffman.
p. cm.—(Proceedings volume / Santa Fe Institute studies in the sciences
and complexity : 9)
Includes bibliographical references and index.
1. Molecular evolution—Congresses. 2. Immune system—Evolution—
Congresses. I. Perelson, Alan S., 1947-. II. Kauffman, Stuart A. III. Title.
IV. Series: Proceedings volume in the Santa Fe Institute studies in the science
of complexity : v. 9.
QH325.W67 1989   574.2'9—dc20   90-1166
ISBN 0-201-52149-0 (HB).—ISBN 0-201-52150-4 (PB)

ISBN 13: 978-0-201-52150-4 (pbk)

This volume was typeset using TEXtures on a Macintosh II computer.

About the Santa Fe Institute

The *Santa Fe Institute* (SFI) is a multidisciplinary graduate research and teaching institution formed to nurture research on complex systems and their simpler elements. A private, independent institution, SFI was founded in 1984. Its primary concern is to focus the tools of traditional scientific disciplines and emerging new computer resources on the problems and opportunities that are involved in the multidisciplinary study of complex systems—those fundamental processes that shape almost every aspect of human life. Understanding complex systems is critical to realizing the full potential of science, and may be expected to yield enormous intellectual and practical benefits.

All titles from the *Santa Fe Institute Studies in the Sciences of Complexity* series will carry this imprint which is based on a Mimbres pottery design (circa A.D. 950–1150), drawn by Betsy Jones.

Santa Fe Institute Studies in the Sciences of Complexity

PROCEEDINGS VOLUMES

Volume	Editor	Title
I	David Pines	Emerging Syntheses in Science, 1987
II	Alan S. Perelson	Theoretical Immunology, Part One, 1988
III	Alan S. Perelson	Theoretical Immunology, Part Two, 1988
IV	Gary D. Doolen et al.	Lattice Gas Methods of Partial Differential Equations, 1989
V	Philip W. Anderson et al.	The Economy as an Evolving Complex System, 1988
VI	Christopher G. Langton	Artificial Life: Proceedings of an Interdisciplinary Workshop on the Synthesis and Simulation of Living Systems, 1988
VII	George I. Bell & Thomas G. Marr	Computers and DNA, 1989
VIII	Wojciech H. Zurek	Complexity, Entropy, and the Physics of Information, 1990
IX	Alan S. Perelson & Stuart A. Kauffman	Molecular Evolution on Rugged Landscapes: Proteins, RNA and the Immune System, 1990

LECTURES VOLUMES

Volume	Editor	Title
I	Daniel L. Stein	Lectures in the Sciences of Complexity, 1988
II	Erica Jen	1989 Lectures in Complex Systems

LECTURE NOTES VOLUMES

Volume	Editor	Title
I	John Hertz et al.	Introduction to the Theory of Neural Computation, 1990
II	Gérard Weisbuch	Complex Systems Dynamics

Contributors to This Volume

C. Amitrano, Dipartimento di Scienze Fisiche, Universita di Napoli, Mostra d'Oltremare, Pad. 19, I-80125, Napoli, Italy

M. Apel, Institute for Genetics, University of Cologne, Weyertal 121, D-5000 Cologne 41, F.R.G.

C. Berek, Institute for Genetics, University of Cologne, Weyertal 121, D-5000 Cologne 41, F.R.G.

Dipak K. Dube, The Joseph Gottstein Memorial Cancer Research Laboratory, Department of Pathology SM-30, University of Washington, Seattle, WA 98195

Herman N. Eisen, Department of biology and Center for Cancer Research, Massachusetts Institute of Technology, Cambridge, MA 02139

Mark I. Greene, Department of Pathology, University of Pennsylvania School of Medicine, Philadelphia, PA 19174

Marshall S. Z. Horwitz, The Joseph Gottstein Memorial Cancer Research Laboratory, Department of Pathology SM-30, University of Washington, Seattle, WA 98195

Gerald F. Joyce, Department of Molecular Biology, Research Institute of Scripps Clinic, 10666 N. Torrey Pines Road, La Jolla, CA 92037

Stuart A. Kauffman, Department of Biophysics and Biochemistry, University of Pennsylvania, Philadelphia, PA 19104-6059, and the Santa Fe Institute, 1120 Canyon Road, Santa Fe, NM 87501 (email: stu@sfi.santafe.edu)

Thomas Kieber-Emmons, IDEC Pharmaceuticals Company, La Jolla, CA

Lawrence A. Loeb, The Joseph Gottstein Memorial Cancer Research Laboratory, Department of Pathology SM-30, University of Washington, Seattle, WA 98195

Catherine Macken, Department of Mathematics, Stanford University, Stanford, CA 94305 (email: cathy@gnomic.stanford.edu)

Wlodek Mandecki, Corporate Molecular Biology D93D, Abbott Laboratories, Abbott Park, IL 60048

Tim Manser, Department of Biology, Princeton University, Princeton, NJ 08544

Richard Palmer, Department of Physics, Duke University, Durham, NC 27706 (email: palmer@physics.phy.duke.edu)

L. Peliti, GNSM-CISM, Unità di Napoli, Associato INFN, Sezione di Napoli, Italy

Alan S. Perelson, Theoretical Division, Los Alamos National Laboratory, Los Alamos, NM 87545 (email: asp@receptor.lanl.gov)

Debra L. Robertson, Department of Molecular Biology, Research Institute of Scripps Clinic, 10666 N. Torrey Pines Road, La Jolla, CA 92037

M. Saber, Laboratoire de Magétisme, Faculté de Sciences, Av. Ibn Battouta, B.P. 1014, Rabat, Morocco

Peter Schuster, Institut für Theoretische Chemie der Universität Wien, Währingerstraβe 17, A-1090 Wien, Austria (email: A8441DAM@AWIUNI11.bitnet)

Gregory W. Siskind, Department of Medicine, Division of Allergy and Immunology, The New York Hospital-Cornell Medical Center, 525 East 68th Street, New York, NY 10021

Contributors to This Volume

Daniel Stein, Department of Physics, University of Arizona, Tucson, AZ 85721 (email: dls@rvax.arizona.edu)

Richard G. Weinand, Department of Computer Science, Wayne State University, Detroit, MI 48202 (email: richard@mercury.cs.wayne.edu)

Edward D. Weinberger, Max Planck Institute, Post Fach 2841, D3400, Göttingen 1, FEDERAL REPUBLIC OF GERMANY

David B. Weiner, Department of Medicine, University of Pennsylvania School of Medicine, Philadelphia, PA 19174

Gérard Weisbuch, Laboratoire de Physique Statistique de l'Ecole Normale Supérieure, 24 rue Lhomond, F 75231, Paris Cedex 5, France (email: weisbuch@FRUMLM62.bitnet)

William V. Williams, Department of Medicine, University of Pennsylvania School of Medicine, Philadelphia, PA 19174

Preface: Applied Molecular Evolution on Rugged Fitness Landscapes

ABOUT THE WORKSHOP

This book grows out of a conference held at the Santa Fe Institute March 27 to 31, 1989. The conference was organized to discuss a central emerging area of biological science: the study of adaptive processes that optimize on rugged fitness landscapes. Since Sewall Wright introduced the concept of fitness landscapes,[9] where each genotype can be thought of as a point in a discrete genotype space and has a fitness, it has become almost second nature for biologists to conceive of adaptive evolution as "hill climbing" towards fitness peaks. Much of the basic research in population genetics, whether concerned with haploid or diploid systems, has involved analysis of the behavior of adapting populations, due to mutation, recombination, and selection, as they flow over fitness landscapes.[4] Classical population genetics has also emphasized frequency and density-dependent selection, coevolution, the evolution of the genetic mechanisms of recombination and sex, and other issues. What then are the new strands which we sought to explore at the Santa Fe Institute meeting?

The dominant answers are three. First, new mathematical tools now permit us to study the *structures* of complex, rugged, multipeaked fitness landscapes. Second, new *experimental techniques* are available to analyze molecular fitness landscapes. Third, current recombinant cloning techniques now allow for the capacity to carry out *applied molecular evolution* to evolve biomolecules of medical and industrial use.

Fitness landscapes underlie both molecular and morphological evolution. Mathematical description of such landscapes can be expected to lead to new experimental studies that actually test and establish the structure of fitness landscapes. Sidestep for the moment the normal biological definition of "fitness." In this book we shall typically be concerned with quite concrete issues where the "fitness" of an enzyme is its capacity to catalyze a given reaction in standard conditions, or the "fitness" of an antibody is its affinity for a defined epitope. Presumably, the fitness or "affinity landscape" of the diverse set of antibodies generated in response to a specific antigen underlies the maturation of the immune response. Thus, much of the book focuses on the actual facts and processes of maturation of the immune response, and their relation to the new theory that is brewing.

Given concrete cases of molecular fitness landscapes, the new theoretical strands include the following questions:

1. What do fitness landscapes underlying molecular or morphological evolution "look like"? How rugged and multipeaked are such landscapes? Are there families of landscapes, where each family ranges from smooth to rugged? How many such families? What are the salient statistical properties of such landscapes? These must include such features as the number of fitness peaks, the lengths of adaptive walks via fitter mutant neighbors to such peaks, the number of mutations accumulated along such adaptive walks, the rate at which the number of directions "uphill" dwindle as successively fitter variants are found, whether or not the global optimum is attainable, how many local optima are accessible from any starting point, whether the local optima are similar to one another or scattered randomly across the space. Other issues include the distribution of fitnesses of optima, and techniques to measure experimentally the correlation structure of a fitness landscape.

2. A second class of questions concerns the *actual flow of an adapting population* over a fitness landscape of a given structure, as a function of the size of the population, the mutation rate, and, if sexual, the mating structure and effects of recombination.

3. A third class of questions about rugged landscapes concerns what kinds of objects, antibodies, enzymes, DNA regulatory sites, or more macroscopic systems such as entire genomic regulatory networks, have what kinds of fitness landscapes and why.

ABOUT THIS BOOK

The first section of this book lays out a number of the general issues concerning the structure of rugged fitness landscapes, with contributions from Palmer; Amitrano,

Peliti, and Saber; Stein; and Schuster. It is not an accident that many of these
workers are solid-state physicists. The expected structures of rugged fitness landscapes is a close cousin to the structure of rugged potential surfaces which arise
in spin glasses and other disordered materials. One purpose of this volume is to
acquaint biological readers with concepts applicable to biological evolution which
have close parallels in statistical physics. These include issues such as the behavior
of an electron in a complex potential surface at a finite temperature, as a function
of the barrier heights between "valleys." Such processes are related to slow relaxation times and "freezing in" in regions of a rugged fitness landscape as a function
of population size and mutation rate.

Rugged, multipeaked landscapes also arise in the protein and nucleic acid folding problems. Here the landscape is a free energy surface and describing movement
on the surface and the question of being trapped at local minima versus the global
minimum arises. In fact, issues of structure underlie the association of a fitness to
a molecular sequence. Thus embedded in the statistical structure of a molecular
fitness landscape is information about the relationship between changes in the underlying sequence and the configuration of a molecule. The fourth chapter in this
volume by Schuster illustrates this point in the context of RNA evolution.

If the structure of fitness landscapes underlies molecular and morphological
evolution, then a central task is to measure and characterize such landscapes. Maturation of the immune response is one favorable system for such an analysis. As
described in detail in many chapters in this volume, the immune response to an
antigen is initiated by stimulation of mitosis in B cells bearing immunoglobulin
receptors of at least modest affinity for the antigen. Activation of B cells is followed
by the induction of a hypermutation mechanism that introduces point mutations
in the binding region of the antibody molecule, the V region, such that some B-cell
variants have reduced affinity for the antigen, others have increased affinity for the
antigen. Presumably the latter are stimulated to divide more, leading to selection of
B cells synthesizing antibodies of successively higher affinity for the antigen. Thus,
maturation of the immune response can be viewed as an adaptive walk in antibody
space from initial B cells harboring receptors with "roughed in" molecular matches
to the incoming fitter mutant neighbors of the antibody molecules in "antibody
space," to or towards high-affinity "peaks" in the fitness landscape.

The second section of the volume, with contributions by experimental immunologists Eisen; Berek and Apel; Manser; Williams, Kieber-Emmons, Weiner, and
Greene; and Siskind, and theoretical biologists Macken and Perelson; Kauffman and
Weinberger; Weisbuch and Perelson; and Weinand, examines both the history and
the current status of experimental work on somatic mutation and the maturation of
the immune response, and discusses current attempts to build detailed mathematical models of the affinity landscape. The chapters by Macken and Perelson, and
Kauffman and Weinberger show how the ideas of rugged landscaped can be used
to understand a number of characteristic features of maturation of the immune
response. Weisbuch and Perelson examine simple models for affinity maturation
in the context of idiotype networks, while Weinand takes a more global view and

presents an intricate computer model of clonal selection and B cell growth based on a three-dimensional representation of *shape-space*.[8]

Study of the structure of fitness landscapes underlying molecular evolution is not merely of scientific interest.[2,3,6,7] It is likely to become of very great practical interest. It is now possible utilizing cloning techniques to generate very large libraries of random, or partially stochastic DNA sequences, hence their corresponding RNA and protein or peptide products. A variety of selection and screening assays to seek products which might have a specific function are under development. Such products may range from new enzymes to new drugs and vaccines. For example, procedures to obtain peptides which mimic an antigen such as insulin might make use of polyclonal or monoclonal antibodies against insulin to screen for partially stochastic peptides bound by those antibodies. Any such peptide is a candidate to harbor an epitope which is similar to the insulin antigen, and hence may mimic it. Such peptides might potentiate, inhibit, or modulate the activity of insulin. If the incoming antigen is the protective antigen on a pathogen, then the mimetic peptide is a candidate vaccine.[2,3,6,7]

Initial work carrying out applied molecular evolution is underway in the laboratories of Greene, Joyce, Kauffman, Loeb, and Mandecki. Patents held by these workers covering many of the initial techniques are pending or have been issued. Ideas and result discussed in the third section by Robertson and Joyce; Mandecki; and Horwitz, Dube, and Loeb.

The final section discusses one of the current serious models of the origin of life, the *hypercycle* model of Eigen and Schuster. This model is of interest in the current setting for at least three reasons. First, Joyce's experiments evolving DNAase activity in a ribozyme is part of the evidence showing that RNA molecules harbor and evolve a variety of catalytic functions. In the hypercycle model, some kind of catalytic coupling of replicating RNA strand pairs around a cycle of such pairs is required. Second, the hypercycle model is an explicit example of *coevolution* in a system of coupled replicating polymers. It therefore bears a relation to idiotype network models where B cells coevolve with one another as a function of affinities of each antibody for its anti-idiotype. Finally, the Eigen and Schuster model is one of several alternatives being explored by members of the Santa Fe Institute community; the others are by Rothkar, Stein and Anderson, Rasmussen and coworkers, Kauffman[7], Farmer et al.[5], and Bagley et al.[1] The latter models were discussed at the meeting, but are not republished here.

ACKNOWLEDGMENTS

This volume, like others in the Santa Fe Institute Series, is devoted to exploring new issues in the sciences of complexity. We hope the volume, like the series, proves useful to a wide readership. The editors are grateful to the Santa Fe Institute for its continuing hospitality, intellectual vigor, and enthusiasm. We are equally grateful to and thank the participants for their efforts, intelligence, and capacity to join minds in a rather new effort. Funding for the workshop came from SFI core funding, including major grants from the MacArthur Foundation, the National Science Foundation (PHY8714918), and the U.S. Department of Energy (ER-FG05-88ER25054).

Stuart A. Kauffman   Alan S. Perelson
University of Pennsylvania   Theoretical Division
& Santa Fe Institute   Los Alamos National Laboratory

July 2, 1990

REFERENCES

1. Bagley, R. J., J. D. Farmer, N. H. Packard, A. S. Perelson, and I. M. Stadnyk. "Modeling Adaptive Biological Systems." *Biosystems* **23** (1989):113–138.

2. The Department of Trade and Industry. "Method for Obtaining DNA, RNA, Peptides, Polypeptides, or Proteins by Means of a DNA Recombination Technique." English patent number 2183661, issued to Marc Ballivet and Stuart Allen Kauffman and dated 6/28/89.

3. Deutsches Patentant. "Verfahren zur herstellung von peptiden, polypeptiden, oder proteinen." German patent number 3,590,766.5-41, issued to Marc Ballivet and Stuart Allen Kauffman and dated 8/13/90.

4. Ewens, W. *Mathematical Population Genetics.* New York, NY: Springer-Verlag, 1979.

5. Farmer, J. D., S. A. Kauffman, and N. H. Packard. "Autocatalytic Replication of Polymers." *Physica* **22D** (1986):50–67.

6. French Patent Office. "Procédé d'obtention d'ADN, ARN, peptides, polypeptides ou proteinés par une technique de recombinaison d'ADN." French patent number 863683, issued to Marc Ballivet and Stuart Allen Kauffman, dated 12/24/87 and registered as 2,579,518.

7. Kauffman, S. A. "Autocatalytic Sets of Proteins." *J. Theoret. Biol.* **119** (1986):1–24.

8. Perelson, A. S. and G. F. Oster. "Theoretical Studies of Clonal Selection: Minimal Antibody Repertoire Size and Reliability of Self–Nonself Discrimination." *J. Theoret. Biol.* **81** (1979):645-670.

9. Wright, S. "The Roles of Mutation, Inbreeding, Crossbreeding, and Selection in Evolution." In *Proceedings of the Sixth International Congress on Genetics* **1** (1932):356–366.

Contents

Applied Molecular Evolution   237

Origin of Life Models   279

Rugged Landscapes
and Evolution

Richard Palmer
Department of Physics, Duke University, Durham NC 27706

Optimization on Rugged Landscapes

Rugged landscapes are a common underlying feature of many complex systems. They are studied from various viewpoints in the physics of glasses and spin glasses, in the biophysics of macromolecules, in the computer science of combinatorial optimization problems, and in the interdisciplinary field of neural networks. In biology there are many potential applications, from pre-biotic evolution to genetic regulatory networks.

This introductory article consists of a review of the notion of a landscape, a discussion of types and properties of landscapes, a description of two models for rugged landscapes, and finally a review of some novel optimization methods for finding the highest or lowest point on a rugged landscape.

LANDSCAPES AND TOPOLOGIES

A *landscape* means simply a single-valued scalar function $F(\mathbf{x})$ of the *state* or *configuration* $\mathbf{x}$ of a system. The variable $\mathbf{x}$ typically has very many dimensions,

Molecular Evolution on Rugged Landscapes, SFI Studies in the Sciences of
Complexity, vol. IX, Eds. A. Perelson and S. Kauffman, Addison-Wesley, 1991

and may usually be written, like a multidimensional vector, in terms of a set of N components x_i:

$$\mathbf{x} = (x_1, x_2, \ldots, x_N).$$ (1)

The term "landscape" comes from visualizing a geographical landscape in which the height h above sea level is a simple function $h = F(x, y)$ of the two-dimensional location $\mathbf{x} = (x, y)$. The rugged landscapes to be considered here normally involve variables $\mathbf{x}$ in many more than two dimensions, but the image is still useful.

For the landscape to be of interest in the context of optimization, the quantity $F(\mathbf{x})$ should be something that we want to maximize or minimize. We assume that $F(\mathbf{x})$ is bounded, so that its value at the maximum or minimum is not infinite. The actual value of F is usually less important than the location $\mathbf{x}_0$ that produces the maximum or minimum. This is sometimes written (for the maximum case)

$$\mathbf{x}_0 = \operatorname*{argmax}_{\mathbf{x}} F(\mathbf{x})$$

whereas $\max_{\mathbf{x}} F(\mathbf{x})$ means the value of $F(\mathbf{x})$ at the maximum.

In biological applications the function $F(\mathbf{x})$ of interest is generically the *fitness* of an individual as a function of its genotype or phenotype. An "individual" here might be a whole organism, or just a small part thereof, such as a particular macromolecule. I entirely shy away from the tricky problem of trying to define "fitness"; it must obviously depend on the environment and on other interacting organisms.

As a particular example we might consider the fitness of a polypeptide chain of length N, based perhaps on its enzymatic efficiency in a particular reaction. The configuration $\mathbf{x}$ would then be the primary amino-acid sequence, which could be represented as in Eq. (1) by N components x_i, each taking one of 20 values. This would give 20^N possible values for $\mathbf{x}$. More realistically, we might consider polypeptides with a range of values for N, in which case Eq. (1) would have to be generalized appropriately.

In other fields similar functions are defined. In physics and chemistry we are often interested in *energy* or *free energy* landscapes, as a function of the positions or orientations of the atoms or molecules making up the system. In computer science one considers the *cost function* or *objective function* for an optimization problem, giving a cost to each particular solution considered. In both these cases we want to *minimize* the function $F(\mathbf{x})$ rather than maximizing it as for biological fitness. This sign inversion—are we maximizing or are we minimizing?—comes up constantly in discussions and in this paper. Unfortunately some of the terms (e.g., basin, lake, hill-climbing) are tied to one viewpoint or the other, so we cannot adopt one choice from the outset.

Strictly speaking, we cannot consider $F(\mathbf{x})$ as a landscape surface unless $\mathbf{x}$ is a continuous variable. This is rarely the case in problems from biology, such as our polypeptide example; there we had a *discrete* set of 20^N possibilities. Nevertheless we can still think in terms of a landscape if we know which values of $\mathbf{x}$ are close to one another. Then we can imagine stepping from point to neighboring point in the *configuration space* of possible $\mathbf{x}$ values and see how the "height" $F(\mathbf{x})$ of the landscape changes as we walk. This notion of closeness among $\mathbf{x}$ values can be supplied by a suitable *metric* $d(\mathbf{x}, \mathbf{x}')$ which gives the "distance" between any two points $\mathbf{x}$ and $\mathbf{x}'$.

For example, in our polypeptide example we could define the distance between two strings (both of length N) to be the number of positions in which the amino acid residues are different. Formally we could write

$$d(\mathbf{x}, \mathbf{x}') = \sum_{i=1}^{N} (1 - \delta_{x_i, x_i'})$$

where the Kronecker delta symbol $\delta_{p,q}$ is defined to be 1 if $p = q$ and 0 otherwise. It is not hard to show that this satisfies the mathematical requirements for a metric. Using this metric the neighbors of a given polypeptide $\mathbf{x}$ are the $19N$ polypeptides that differ by a single residue; we call these the *one-mutant neighbors* of the original polypeptide.

The choice of a metric is not necessarily unique, and in some cases quite different topologies can be placed on the same problem. This *is* a matter of significance, because most practical optimization algorithms work in terms of gradual improvement by making steps that are short in the chosen metric. Some metrics may make the landscape appear much more rugged than others. A good metric is usually one in which the landscape height changes only slowly as one moves from point to neighboring point; $F(\mathbf{x})$ and $F(\mathbf{x}')$ should not differ greatly when $d(\mathbf{x}, \mathbf{x}')$ is small.

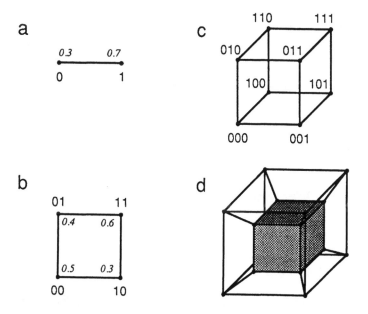

FIGURE 1   Configuration spaces for binary variables: (a) $N = 1$; (b) $N = 2$; (c) $N = 3$; (d) $N = 4$.

The chosen metric defines the *topology* of the configuration space in which **x** lives. "Topology" in this context means essentially *what is connected to what*. Figures 1 and 2 show some examples of simple configuration spaces for discrete variables. Figure 1 is for binary-valued components, $x_i = 0$ or 1, in $N = 1$, 2, 3, and 4 dimensions. The vertices represent the 2^N values of **x**, while the lines join neighbors. In the first three cases ($N = 1$ to 3) the vertices are labeled by coordinates $x_1 \ldots x_N$ in an obvious way. In the first two cases we have also shown some possible fitness values for each vertex. Note that in case (b) there are two *local* maxima—points that are more fit than all their one-mutant neighbors—at 00 and 11, but only one *global* maximum, at 11.

Figure 2 is for ternary (three-valued) variables x_i with $N = 1$ in (a) and (b) and $N = 2$ in (c) and (d). We see that two distinct topologies are natural, depending on whether each three-valued component x_i is cyclic or linear; in the cyclic case values 0 and 2 are just as close as 0 and 1 or 1 and 2, while in the linear case you can only get from 0 to 2 by going through the intermediate value 1. This is simple for $N = 1$ in (a) and (b), but much more complicated for $N = 2$ in (c) and (d). It is interesting to try to visualize the corresponding configuration space for $N \geq 3$; only the linear case is easy, and then only for $N = 3$.

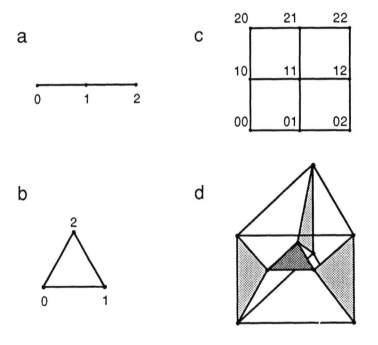

FIGURE 2 Configuration spaces for ternary variables: (a) $N = 1$, linear; (b) $N = 1$, cyclic; (c) $N = 2$, linear; (d) $N = 2$, cyclic.

For systems in which there are more than 3 values for each x_i there are correspondingly more possible topologies that might reasonably arise. In the polypeptide example, at a given site we might regard the 19 alternative amino-acid residues as equidistant from a given one, as assumed above, or we might have a more sophisticated criterion of similarity which would lead us to designate only *some* of these single mutants as neighbors. Similarity might be defined in terms of similar fitness, giving a smoother landscape surface, or on the basis of physical or chemical substitutability.

More generally, the notions of distance and neighborhood do *not* necessarily factorize into a sum over components of the form

$$d(\mathbf{x}, \mathbf{x}') = \sum_{i=1}^{N} d_i(x_i, x_i') \, .$$

For example, the effective difference between lysine and arginine at one site might depend crucially on what is at another site 50 or 100 residues away on the primary sequence. Such effects are usually described solely by the fitness function; it is impractical to construct an appropriate topology to describe them in the connectivity of the underlying configuration space.

TYPES OF LANDSCAPES

It is worth drawing some simple pictures to illustrate different types of landscapes. Of course it is impossible to be realistic for a many-dimensional configuration space, so we normally content ourselves with using one or two continuous variables to represent the state $\mathbf{x}$, and plot the fitness function $F(x)$ or $F(x, y)$ upwards. This can be very deceptive but is better than no picture at all. In each of the following cases we discuss particularly the problem of finding the location of the maximum, starting from an arbitrary point on the landscape and walking continuously on it.

If our landscape resembles Figure 3(a) the maximization problem is easy and uninteresting. We can simply climb up the surface to the global maximum. This is equally true in many dimensions, as evidenced by the two-dimensional case in Figure 3(b). Any rule that always takes an uphill direction will find the maximum, although we can help it to get there rapidly by going in the gradient direction. In a continuous space we can also optimize how far we move at each step, making a compromise between an excruciatingly slow approach and the danger of overshooting. A common ploy, called *gradient ascent* or *hill climbing*, is to take steps of a size (as well as direction) proportional to the gradient, so that the step size decreases near the top. There are also much more sophisticated approaches available, such as *conjugate gradient methods*.[33] In general these can speed up the maximization process immensely, particularly when the maximum is much broader in some directions than in others, as in Figure 3(c). In such cases a naive gradient ascent algorithm is liable to reach the ridge rapidly but then take a long time to follow it to the top, zigging and zagging from side to side if the step size is large.

On the other hand a single maximum is not necessarily easy to find if the landscape is not smooth. Extreme cases in one and two dimensions are shown in Figure 4, where there are *no* gradient clues as to the location of the summit. The only way to find it is to search the whole configuration space. This makes the problem extremely hard in high dimensions, and indeed some NP-complete[10] problems can be formulated in terms of such a landscape.[4] In the physics literature the equivalent energy surface with one low point in an otherwise flat landscape is often called a *golf-course potential*.

Figure 5 shows several cases in which there are two or more *equivalent* maxima. This situation is common in physics, where the maxima (or minima) are related by symmetries in the problem. The maximization problem is again easy, since climbing up to the nearest maximum from any starting point will provide a solution as good as any other. Such a problem is said to exhibit *symmetry breaking* because the solution does not possess the full symmetry of the problem. It also represents a *lock-in* phenomenon, in which a system becomes stuck in one of several states depending on initial conditions.

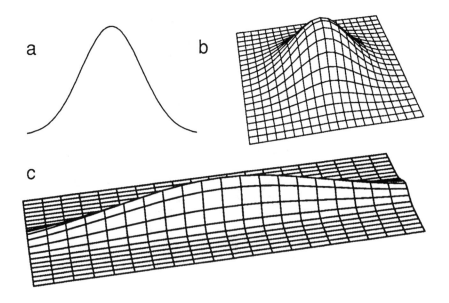

FIGURE 3 Landscapes with a single maximum.

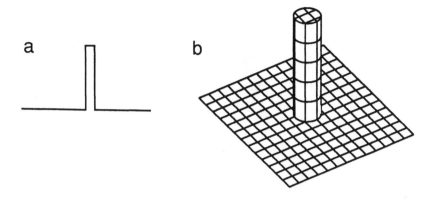

FIGURE 4 Flat landscapes with an isolated single maximum.

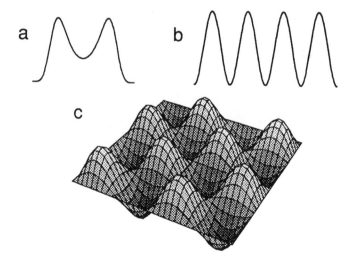

FIGURE 5 Landscapes with many equivalent maxima.

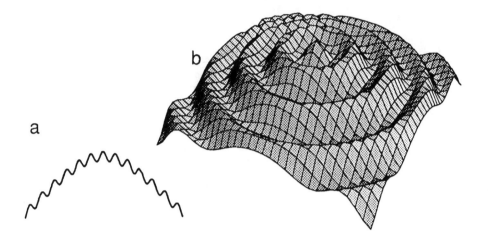

FIGURE 6   Landscapes with many local maxima and one global maximum.

Less straightforward problems are shown in Figure 6. Here, although there is only one global optimum, there are many local optima in which one can become stuck. Continuous gradient ascent will obviously fail, but it could succeed if used with discrete steps large enough to get past the fine structure. Even so, there are tricky questions about the optimum step size, or about how to design an algorithm with a self-adjusting step size. A better basis would be an algorithm that is able to accept some jumps that *decrease* fitness somewhat, while preferring jumps that give an increase. This is the essence of the simulated annealing method discussed later.

Finally, Figure 7 represents some *rugged* landscapes, with many irregular local maxima. There are typically—at least in the cases of particular interest—a large number of roughly equivalent maxima with fitness close to the optimum, but these are *not* related by symmetry. The landscape may even be self-similar, so that a magnified image of a part would be much like the whole surface, with structure on every scale. An algorithm to find one of the high maxima must be able to take long jumps, or to descend to lower fitness levels at least occasionally, and must not be satisfied with *any* local maximum until considerably more exploration has taken place around it.

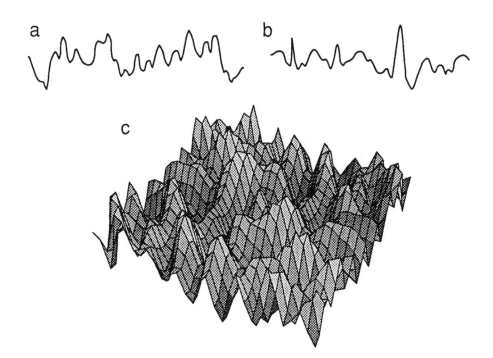

FIGURE 7   Rugged landscapes in one and two dimensions.

PROPERTIES OF LANDSCAPES

There are many ways to characterize rugged landscapes and to distinguish classes within them. Various statistical properties are useful both for this classification and for applications. We list some of the more interesting ones.

1. Number and fitness distribution of local maxima. Usually we would look at how the number N_{max} of local maxima scales with the size N of the problem. In most cases of interest the scaling is at least as fast as exponential, typically $N_{max} \propto \exp(aN)$ or $N_{max} \propto N!$ (factorial). Indeed this at-least-exponential scaling is sometimes used to define what is meant by a *rugged* landscape or a *complex* system. For more focused information we might count just those maxima that lie within (say) 5% of the global maximum. Most generally we might hope to construct a distribution function $g(F)$ (often called a *density of states* function) so that $g(F)df$ gives the number of local maxima in the fitness interval $[F, F + df]$.

2. Distribution of barriers. The paths between maxima may also be important. In physics the "barrier heights" of the lowest paths between pairs of minima (recall the inversion!) govern the likelihood of a transition between the physical states that the minima represent. In any application of simulated annealing it is important to know how far one needs to back off from the local maxima or minima to find transition paths between them; this sets the scale of the temperature parameter. Again we can distinguish different landscapes by the way in which the barrier heights or back-off distances scale with N. A power law, height $\propto N^\alpha$, is often found.

3. Number of "lakes" as a function of flood-level. Consider a physicist's energy landscape with lower points more probable. It is often useful to consider what happens if we "flood" the landscape to a given depth, to represent permitting (through thermal excitation) all energies up to the given value. For most depths we will find a set of disjoint *lakes*, each of which represents one region of configuration space in which the system can become stuck. As we raise the flood-level these lakes will necessarily merge in a hierarchical manner,[36,38] as illustrated in Figure 8. The resulting hierarchical tree is a crucial ingredient for understanding the physics of many complex systems.[2,28,29,32] It may itself be characterized by various statistical measures. For some applications we may also need to know the volume V (of $\mathbf{x}$ space) in each lake, or an average across all lakes. In physics these volumes are directly related to the entropy S of the corresponding thermodynamic state via Boltzmann's relation $S = k_B \log V$.

4. Correlations between height and width of maxima. To what extent is a high maximum on top of a wide mountain? Figures 7(a) and 7(b) contrast cases in which high maxima have wide (7a) and narrow (7b) basins. Or, looking for minima, to what extent are the deepest points beneath the widest lakes? It is clear that narrow minima are much harder to find than broad ones. Techniques such as simulated annealing spend more time in the wider lakes, and so naturally find the deepest minima as the flood-level (temperature) is lowered *if* width is well correlated with depth.

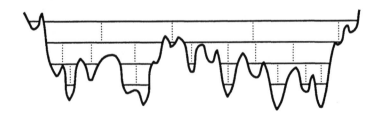

FIGURE 8 A rugged landscape shown flooded to four different levels. The lakes merge hierarchically as the flood-level increases.

5. Correlations between locations of good optima. Suppose that we know about the locations $\mathbf{x}^{(1)}$ and $\mathbf{x}^{(2)}$ of two deep minima. Where else might we look for even deeper ones? This obviously depends on the structure of the landscape, but there are several extreme cases worth mentioning. First, in the worst case, it might be that knowing $\mathbf{x}^{(1)}$ and $\mathbf{x}^{(2)}$ gives *no* information about further minima, as is obviously true on a fully random landscape. Second, it might be a good idea to look along the line $\alpha\mathbf{x}^{(1)} + (1 - \alpha)\mathbf{x}^{(2)}$ joining $\mathbf{x}^{(1)}$ and $\mathbf{x}^{(2)}$. Third, it might be useful to construct new $\mathbf{x}$'s using some components from $\mathbf{x}^{(1)}$ and some from $\mathbf{x}^{(2)}$:

$$\mathbf{x} = \left(x_1^{(n_1)}, x_2^{(n_2)}, \ldots, x_N^{(n_N)} \right)$$

where $n_i = 1$ or 2 for each i. This is the basis for the *genetic algorithms* that will be discussed later. It is particularly useful in situations where the fitness function $F(\mathbf{x})$ approximately factorizes into a sum (or product) of terms in each coordinate,

$$F(\mathbf{x}) \approx \sum_i F_i(x_i) . \tag{2}$$

This is called an *additive fitness model* in biology.

6. Properties of walks. We can define various types of walks on landscapes, such as those that take randomly any upward direction at each step, or those that take the direction that goes up most steeply. For a given type of walk, starting from a random starting point, we may ask for a variety of statistical properties. What is the average length of a walk before a maximum is reached? How does the number of available (uphill) directions decrease as the walk proceeds? How many maxima are accessible through walks of a given type? Stuart Kauffman and his colleagues have answered many such questions for walks on the NK landscapes described below.[17]

MODELS OF LANDSCAPES

Considerable progress has been made on the problem of understanding rugged landscapes through the use of simple models. Even where fully known, the real landscapes of interest are usually too complicated to deal with effectively, and must be simplified or modeled. Simple formal models are easier to deal with, both analytically and computationally, and may help us to develop appropriate intuitions for dealing with more realistic cases.

We may speculate that some landscape properties are independent of many details of the landscape. For example, some of the properties of walks mentioned in 6. above depend only on the rank ordering of fitnesses, not on the specific fitness values. This idea of generic behavior independent of model details is common in physics,

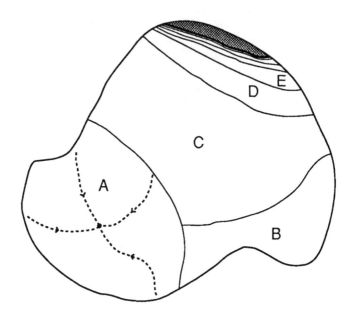

FIGURE 9 A parameter space divided into several universality classes. Different points correspond to different sets of parameters describing the system. Renormalization group trajectories leading to a fixed point are sketched in universality class A. Universality breaks down at the limit of the sequence C, D, E, . . .

where we find that the *parameter space* of a wide range of models breaks up into just a few *universality classes*, as shown in Figure 9. Within each universality class many properties are generic, and we have to change the parameters greatly to switch into another class. In physics we understand this phenomenon through the ideas of the *renormalization group*, which provides trajectories in parameter space along which properties are generic; this divides up parameter space into the basins of attraction of the different *fixed-points* of these trajectories. No analogue of the renormalization group has been suggested for rugged landscapes outside physics, but we may reasonably expect a similar decomposition into universality classes. If that is the case, then the simplest model within a universality class will suffice for determining and describing the generic properties of that class; the details of the real system may not matter as long as we are in the right universality class. Note, however, that this attitude can sometimes be taken too far and act as an excuse for *over*-idealization.

It is useful to describe the two most prominent models, Stuart Kauffman's NK model, and the SK spin glass model. These do not exhaust the possibilities, and there is a need for some even simpler rugged landscape models amenable to analytic mathematical analysis.

THE NK MODEL

Kauffman[16] introduced the NK model to describe the complex genetic regulatory systems. It has since been applied to many rugged landscape problems. Only a brief description is given here; the interested reader should consult Kauffman's articles for details.[17,18,19]

Consider a system of N binary variables $x_i = \pm 1$, so that $\mathbf{x} = (x_1, x_2, \ldots, x_N)$ is a vertex of an N-dimensional hypercube. We construct an overall fitness function as a sum of N contributions, with the ith contribution depending on x_i and K other x_j's:

$$F(\mathbf{x}) = \sum_{i=1}^{N} F_i(x_i, x_{j_1}, x_{j_2}, \ldots, x_{j_K}).$$

The dependence of F_i on the K x_j's represents epistatic interactions between the allele at locus i and those at loci $x_{j_1} - x_{j_K}$. The particular indices j_k depend on i. They may be chosen randomly for each i and k (normally excluding repeats for the same i), or may for example label K near neighbors of x_i when the variables are arranged on a string or lattice. The functions $F_i(x_i, \ldots, x_{j_K})$ are normally random functions of their arguments, with values chosen from a probability distribution $P(F_i)$ independently for each of the 2^{K+1} possible values of the arguments.

Once all these random choices have been made they are frozen. Then $F(\mathbf{x})$ is a well-defined single-valued function of $\mathbf{x}$ whose landscape properties may be investigated. This has been carried out mainly though computer simulation, though some simple properties may also be derived mathematically.

Two extreme cases are of special interest. If $K = 0$ we have simply

$$F(\mathbf{x}) = \sum_{i=1}^{N} F_i(x_i)$$

which is an additive fitness model. It has a single global maximum (barring accidental degeneracy) and no local maxima, as in Figure 3. It is easy to find the maximum; we simply maximize each variable x_i separately, in any order. Since each $F_i(x_i)$ takes only two values, we may write

$$F_i(x_i) = a_i + b_i x_i$$

and hence

$$F(\mathbf{x}) = A + \sum_{i=1}^{N} b_i x_i. \tag{3}$$

In physics this would be called a random field paramagnet.

At the other extreme consider $K = N - 1$. Now every variable appears in every F_i, and thus $F(\mathbf{x})$ is the sum of N independent random numbers, independently for each possible state $\mathbf{x}$. The landscape is thus totally random, each point being uncorrelated with all others, and there is clearly no technique better than exhaustive search for finding the global maximum. For large N the central limit theorem implies that the individual points have a Gaussian fitness distribution.

As K increases from 0 to $N-1$ the landscape goes from having a single smooth maximum to being totally random. Between these limits it becomes increasingly rugged, with more and more local maxima and less and less local correlation. This is the great advantage of the NK model; the K parameter may be used to *tune* the ruggedness as desired.

THE SK SPIN GLASS MODEL

The materials called spin glasses are mainly alloys of a magnetic and a non-magnetic metal, such as a few percent of iron dissolved in gold.[6,8,37] The metals can be mixed while molten and then cooled to form a crystal—a solid with the atoms arranged on a periodic lattice. The magnetic atoms are called spins, and take up random positions on the lattice. Pairs of spins interact with one another mainly via polarization of the electrons in the metal, in a way that depends crucially on the distance between the two spins concerned. Some pairs of spins have lower energy when parallel, others when anti-parallel. The problem of minimizing the *total* energy, from all pairs of spins, is highly non-trivial; the energy function has many local minima and is generally very rugged.

A simple idealized model, called the Sherrington-Kirkpatrick[34] or SK model, represents each spin by a binary (or *Ising*) variable $x_i = \pm 1$ (for $i = 1, 2, \ldots, N$), in effect allowing only two directions for each. Each spin interacts with *all* others (even though this is very unrealistic), with an energy contribution of the form $-J_{ij} x_i x_j$ where J_{ij} is a random coefficient that may be positive or negative. The total energy (or *Hamiltonian*) is

$$E(\mathbf{x}) = -\sum_{ij} J_{ij} x_i x_j .$$

Note that this is closely related to an NK model with $K = 1$, which could be written

$$E(\mathbf{x}) = -\sum_{ij} [a_i + b_i x_i + c_{ij} x_i x_j]$$

by analogy with Eq. (3), but where the sum would normally be taken over only N selected pairs instead of over all $N(N-1)/2$ pairs.

The SK model has been studied in great detail,[27] and several new techniques, such as the *replica method*, have been developed for its analysis. Some of these are more widely applicable and have been carried over into other domains.[1] For instance, combinatorial optimization problems (e.g., the traveling salesman problem), can be treated with the replica method.[26]

The NK model and the SK spin glass model may both be used as formal models of rugged landscapes. Their further study may well lead to a deepened understanding of rugged landscapes in general. The main differences between them are:

1. The NK model has tunable ruggedness; the spin glass model does not.
2. The NK model has some biological plausibility in terms of epistatic interactions. The spin glass model was designed for magnetic alloys and has little direct biological plausibility.
3. The NK model is hard to analyze except by simulation. The spin glass model is easier to analyze mathematically, and many tools have been developed for its study.

OPTIMIZATION METHODS

I turn finally to look at some interesting optimization methods appropriate for rugged landscapes. I make no attempt to be complete, especially as there are many books on standard methods. Instead, the focus is on a few novel methods from physics, biology, and computer science.

SIMULATED ANNEALING

In physics and elsewhere complicated systems have long been studied using *Monte Carlo simulation*.[5] The idea is to construct a trajectory $\mathbf{x}(t)$ through the configuration space such that on average particular points are visited with frequency proportional to their Boltzmann-Gibbs probability,

$$\mathrm{Prob}(\mathbf{x}) \propto \exp\left(-\frac{E(\mathbf{x})}{k_B T}\right) . \tag{4}$$

Here $E(\mathbf{x})$ is the energy of configuration $\mathbf{x}$, and is a rugged landscape in the cases of interest. T is temperature of the system and k_B is Boltzmann's constant. As $T \to 0$ the probability becomes negligible for anything but the lowest energy state, or *ground state*, so we have in principle a way of finding the minimum of $E(\mathbf{x})$.

The trajectory $\mathbf{x}(t)$ is constructed from a random starting point by a simple algorithm, repeated again and again[25]:

1. From the current point $\mathbf{x}$, pick a possible move to a neighboring point $\mathbf{x}'$.
2. If $E(\mathbf{x}') < E(\mathbf{x})$, accept the move and make $\mathbf{x}'$ the current point.
3. Otherwise, compute $p = \exp(-[E(\mathbf{x}') - E(\mathbf{x})]/k_B T)$ and accept the move with probability p or reject it (keeping $\mathbf{x}$ as the current point) with probability $1-p$.

It is not hard to show that this leads to Eq. (4) if all states are accessible from the starting point and if we run the algorithm for long enough. Initial parts of the trajectory, before we reach *equilibrium*, must be discounted. Unfortunately, the computer time $t_{\max}$ required to get to equilibrium may be prohibitively long. If the algorithm is not run for a long enough time, one obtains averages over only part of the configuration space, often corresponding to a metastable state of the system. At small T the problem becomes severe, and the algorithm can become stuck for extremely long times in a local minimum of the energy surface $E(\mathbf{x})$. Not surprisingly this situation is worst with very rugged landscapes, where there are very many local minima.

If one specifically wants to find the ground state $\mathbf{x}_0$, it is little use running the Monte Carlo algorithm at $T = 0$, where it only takes downhill steps. A better approach is *simulated annealing*,[20,21] in which one starts at high T (where it is easy to run for an adequate $t_{\max}$) and then gradually lowers T towards zero according to an *annealing schedule* $T(t)$. Information from the current averages can be used to adjust the annealing schedule so that more time is spent in certain temperature regimes where, for example, many components x_i are freezing out. A perfect answer is only obtained with infinitely slow annealing, but a practical schedule can give reasonable results, typically within a few percent of the global minimum.

Simulated annealing can obviously be applied to many types of optimization problems as well as to statistical mechanics problems. One introduces a pseudo-temperature T and gradually lowers it while running a Monte Carlo simulation. At first the typical states on the trajectory are not very good, and are more representa-tive of typical $\mathbf{x}$'s than of optimum ones. As T is lowered the fitness becomes more important, and the $\mathbf{x}$'s on the trajectory move on average towards lower energy (higher fitness). Eventually only points close to an optimum are generated in the trajectory, with finer discrimination appearing as T approaches zero.

Simulated annealing works reasonably well if the low minima occur in the widest valleys, which seems to be the case for at least some of the examples that have been studied. The performance of the algorithm is rarely as good as the most cleverly designed heuristic algorithms for the same problem, but it is often easier to implement because a detailed analysis is not needed. It *is* generally much better than naive algorithms that just do local optimization. Some industrial applications have been found useful, particularly in circuit design and placement.[35]

TEMPORARY INSANITY

Approaches like simulated annealing can often be enhanced by temporarily extend-ing the configuration space, even allowing states that make no physical sense. By relaxing certain constraints on the physical situation the algorithm can explore more widely, and less easily become stuck in a poor local minimum. The effect is to connect regions or valleys that were previously only accessible via large barriers. We may actually change the definition of "neighborhood" in the underlying config-uration space, or may simply lower the energy (or increase the fitness) of certain transitional points. We may even add new points to the configuration space. If the added states or passages are physically unacceptable, they can be discouraged by adding a *penalty function* to the original energy $E_0(\mathbf{x})$:

$$E(\mathbf{x}) = E_0(\mathbf{x}) + \lambda G(\mathbf{x}).$$

The penalty function $G(\mathbf{x})$ is chosen to be large (we are minimizing here, so large means less likely) when the configuration is unacceptable, and small otherwise. The strength λ of the penalty can be adjusted as a function of T to exclude the unacceptable states progressively as annealing proceeds. Such penalty functions must be treated very carefully however, and can give rise to false minima.[9,23]

The classic example is that of placing integrated circuit chips on a circuit board, trying to find the arrangement that minimizes the wiring length between the chips.[21] It is not hard to construct a fitness function depending on the chip locations, and then try to maximize it with simulated annealing. When the allowed moves are simply small displacements of chips in the plane, one rapidly gets stuck in a local minimum, because the chips get in each other's way. On the other hand, if we relax the obvious constraint that chips cannot overlap or occupy the same space, then they can pass through each other and explore many more possibilities. Of course we have to add a penalty function to penalize overlapping chips, and eventually turn up λ so as to inhibit overlapping altogether, but this does not negate the strong effect of allowing overlaps in the earlier stages of annealing.

NEURAL NETWORKS

Artificial neural networks are currently under intensive study in many disciplines.[12] They consist of a large number of individual units (or *neurons*), each receiving input from many others. The output of a unit (which is sent on to others) is a function of its inputs. The connections (or *synapses*) have strengths that may be changed to alter the function of the network. In most cases the connection strengths are adjusted gradually in an automatic way as the network "learns" to solve a problem or represent a relationship. In other cases, including most applications to optimization problems, the network has connection strengths that are pre-designed for a particular purpose.

The most well-known example is Hopfield and Tank's[14,15] neural network to solve the traveling salesman problem (TSP). In the TSP we have N cities $i = 1, 2, \ldots, N$ with known distances d_{ij} between them. The task is to find the minimum-length closed tour that visits each city once and returns to its starting point. This is an NP-complete problem.[10]

We choose N^2 binary units n_{ia} which are always either on (1) or off (0). A particular state of these units represents a particular tour: unit n_{ia} is on if and only if city i is the ath stop on the tour. The total length of the tour is

$$L = \tfrac{1}{2} \sum_{ij,a} d_{ij} n_{ia}(n_{j,a+1} + n_{j,a-1})$$

and there are two constraints:

$$\sum_a n_{ia} = 1 \qquad \text{(for every city } i\text{)}$$

and

$$\sum_i n_{ia} = 1 \qquad \text{(for every stop } a\text{).}$$

The first constraint says that each city appears only once on the tour; the second says that each stop on the tour is at just one city.

We can construct an energy function by adding to the length L penalty terms which are minimized when the constraints are satisfied:

$$E = \tfrac{1}{2} \sum_{ij,a} d_{ij} n_{ia}(n_{j,a+1} + n_{j,a-1}) + \frac{\lambda}{2} \left[\sum_a \left(1 - \sum_i n_{ia}\right)^2 + \sum_i \left(1 - \sum_a n_{ia}\right)^2 \right].$$

Consider the contribution to this expression from a particular unit n_{ia}. This can be written in the form

$$E_{ia} = -n_{ia}\left[v_{ia} + \sum_{jb} w_{ia,jb} n_{jb}\right]$$

where v_{ia} and $w_{ia,jb}$ are constants. We can build (or simulate) physical connections in our neural network so that unit n_{ia} receives as net input the required coefficient $[v_{ia} + \sum_{jb} w_{ia,jb} n_{jb}]$. Now—the central result—the energy is always lowered if a unit turns on when it has a positive input and off when it has a negative input.

We can let each unit turn on and off according to this rule, selecting them in random order to do the updating. This leads in time to a local minimum of the energy function. We are walking downhill on the rugged energy landscape, but with a network in which the computation is spread out over N^2 parallel units.

For a useful solution closer to the *global* minimum we can do one of two things:

1. Use *stochastic* units which sometimes choose the "wrong" value of n_{ia}, according to a temperature parameter T as in Eq. (4). Then we gradually lower T as for simulated annealing.
2. Use continuous-valued units, so that n_{ia} can be anywhere between 0 and 1 (according to its input), and add further penalty terms to force $n_{ia} = 0$ or 1 eventually. Adjust the penalty coefficients as the calculation proceeds.

In either case the system explores many solutions (and non-solutions, with the constraints not strictly true), and finally settles close to the global minimum.

The results of these approaches are good, but not as good as the best ones obtained on the problem by conventional algorithms.[24] The system still has trouble getting caught in local minima of the energy function. But neural networks for a particular computation are often rather easy to design and build (or simulate), whereas good heuristic methods usually only appear after much experience.

Other algorithms can also be implemented in neural networks. For the traveling salesman problem the *rubber band method* of Durbin and Willshaw[7] is particularly successful. See Hertz, Krogh, and Palmer[12] for further discussion.

STATISTICAL MECHANICS

The traditional tools of statistical mechanics, and even the new ones developed for complex systems, do not let us optimize on a *particular* rugged landscape. They can normally only be applied to calculating averages over possible problems drawn from some distribution, not to any particular instance of the problem. The averages can be quite useful though, and tell us for example how well on average we can expect to do in numerical optimization, how the optimum fitness scales with the problem size, and how many solutions to expect. Some of the properties of landscapes listed earlier can be calculated analytically by these methods.

Among rugged landscape problems statistical mechanics approaches have been applied mainly to spin glasses and to a few combinatorial optimization problems such as the traveling salesman problem.[3,9,22,26] Particularly useful has been the replica method. For an overview of this field see Palmer.[30]

GENETIC ALGORITHMS

A very interesting approach to numerical optimization was introduced by Holland[13] in the middle 1970's but is only recently becoming well known.[11] The central idea is to work with a whole *population* of "good" points $x^{(k)}$ instead of just one. There are various ways we may improve the fitness of the points individually, such as simple gradient ascent or simulated annealing. But we can also use information from the population as a whole. To use the biological analogy, we might think of the different $x^{(k)}$'s as the genotypes of a collection of individuals. By mating them in pairs and selecting the fittest, we can hope to combine the best genes of both parents in selected offspring. In terms of the components x_i, it might for example be the case that $x_1^{(23)}$ and $x_2^{(38)}$ are particularly good values for x_1 and x_2 respectively; by mating individuals 23 and 38 we can hope to bring them together into a single x. The general idea is that good components or *building blocks* can evolve independently and then combine. Of course we are here assuming that the overall fitness is to some extent an additive function of its parts, as in Eq. (2), but the technique does not require a strictly additive fitness model.

In more detail, suppose we keep a population of size K. Let us represent each member $x^{(k)}$ (with $k = 1, 2, \ldots, N$) by a binary string $x_1 x_2 x_3 \ldots x_N$ where $x_i = 0$ or 1. Each string has an associated fitness $F^{(k)} = F(x^{(k)})$. We work in discrete time, and construct the population at time $t + 1$ from that at time t according to the following steps:

1. Select two members of the old population, $x^{(a)}$ and $x^{(b)}$ say, each with probability according to its fitness $F^{(k)}$. Selection is done with replacement. The fitnesses may be scaled before use.

2. With probability p_c perform a crossover operation on the two chosen strings. To do this, pick a random number k in the range 0 to N and construct one new string from the first k bits of $x^{(a)}$ and the last $N - k$ bits of $x^{(b)}$, and the other new string in the reverse way.

3. Go through each bit of the two new strings and flip each with (very small) probability p_m. This is a mutation operation that is needed to introduce new diversity into the population.

4. Go back to step 1 and repeat until K new strings have been generated.

This procedure produces many bad strings as well as good ones, but the selection step (1) preferentially keeps the good ones. There are many variations on the basic scheme, but the above serves as an example.

After many generations the population should end up clustered around the global maximum of the fitness function. In practice the technique is not a panacea, but sometimes works well. It is essentially a search technique—note that no derivatives, nor even continuity, are needed—that works on binary strings and tends to keep nearby bits together. It is thus best suited to problems where the correlations between bits (as reflected in fitness) can suggest an appropriate linear ordering, although there are some techniques for automatically re-ordering the bits.[11]

Genetic algorithms are potentially very powerful, but have not yet been sufficiently explored. Most of the practitioners tend to apply what has worked in the past to new problems with very little analysis. Few comparative studies have been carried out. Very little has been done on combining genetic algorithm search techniques with other methods such as simulated annealing, conjugate gradient techniques, and so on. It is clear that genetic algorithms will not always perform well (e.g., for fitness functions that don't approximately factorize), and it is important to define more clearly their efficiency compared to other techniques.

BEYOND RUGGED LANDSCAPES

It would be a mistake to leave the impression that the problems in complex systems can all be reduced to the study of rugged landscapes. While rugged landscapes are indeed a focus of much current work, not all systems of interest can be represented so simply.

Firstly, there are of course many systems with no single fitness (Lyapunov) function. Nevertheless one may still examine the configuration space geometry and/or dynamics, as emphasized in the first section. In some cases, such as the physics of glasses, one can learn a great deal about the dynamic behavior just from the topology of configuration space; steric hindrance of the atoms leaves only tortuous labyrinthine paths between points in configuration space.[31] In general the number and kind of paths between multiple equilibria may be as important as the multiple equilibria themselves.

Secondly, effort is just beginning on examining *co-evolution* of coupled systems where different sub-systems have their own adaptive goals, usually conflicting. The evolution or adaptation of one organism on its landscape changes the landscape of another organism, and vice versa. This is going to be vital in the future as we come to grips with the forefront problems of biology, economics, and the social sciences.

Thirdly, there is the possibility of landscape design, as opposed to working with a given fixed landscape. In some of the current neural network models the process of training the network involves designing a landscape with minima at pre-specified locations. In the co-evolutionary context we will have to deal with one sub-system trying to change, or understand, or mimic, the landscape of another, building up in some way an internal representation of its exterior world.[18]

REFERENCES

1. Anderson, P. W. "Spin Glass Hamiltonians: A Bridge Between Biology, Statistical Mechanics and Computer Science." In *Emerging Syntheses in Science*, edited by D. Pines. Reading, MA: Addison-Wesley, 1986, 17–20.
2. Bachas, C. P., and B. A. Huberman. "Complexity and the Relaxation of Hierarchical Structures." *Phys. Rev. Lett.* **57** (1986):1965–1969, 2877.
3. Baskaran, G., Y. Fu, and P. W. Anderson. "On the Statistical Mechanics of the Traveling Salesman Problem." *J. Stat. Phys.* **45** (1987):1–25.
4. Baum, E. "Intractable Computations without Local Minima." *Phys. Rev. Lett.* **57** (1986):2764-2767.
5. Binder, K., editor. *Monte Carlo Methods in Statistical Physics*. Berlin: Springer, 1979.
6. Binder, K., and A. P. Young. "Spin Glasses: Experimental Facts, Theoretical Concepts, and Open Questions." *Rev. Mod. Phys.* **58** (1986):801–976.
7. Durbin, R., and D. Willshaw. "An Analogue Approach, to the Traveling Salesman Problem Using an Elastic Net Method." *Nature* **326** (1987):689–691.
8. Fischer, K. H., and J. A. Hertz. *Spin Glasses*. Cambridge: Cambridge University Press, 1990.
9. Fu, Y., and P. W. Anderson. "Application of Statistical Mechanics to NP-Complete Problems in Combinatorial Optimization." *J. Phys. A* **19** (1986):1605–1620.
10. Garey, M. R., and D. S. Johnson. *Computers and Intractability: A Guide to NP-Completeness*. New York: Freeman, 1979.
11. Goldberg, D. E. *Genetic Algorithms in Search, Optimization & Machine Learning*. Reading, MA: Addison-Wesley, 1989.
12. Hertz, J. A., A. S. Krogh, and R. G. Palmer. *An Introduction to the Theory of Neural Computation*. Redwood City, CA: Addison-Wesley, 1990.
13. Holland, J. H. *Adaptation in Natural and Artificial Systems*. Ann Arbor: University of Michigan Press, 1975.
14. Hopfield, J. J., and D. W. Tank. "'Neural' Computation of Decisions in Optimization Problems." *Biol. Cybernet.* **52** (1985):141–152.
15. Hopfield, J. J., and D. W. Tank. "Computing with Neural Circuits: A Model." *Science* **233** (1986):625–633.
16. Kauffman, S. A. "Homeostasis and Differentiation in Random Genetic Control Networks." *Nature* **224** (1969):177–178.
17. Kauffman, S. A. "Adaptation on Rugged Fitness Landscapes." In *Lectures in the Sciences of Complexity*, edited by D. L. Stein. Redwood City, CA: Addison Wesley, 1989, 527–618.
18. Kauffman, S. A. "Principles of Adaptation in Complex Systems." In *Lectures in the Sciences of Complexity*, edited by D. L. Stein. Redwood City, CA: Addison Wesley, 1989, 619–712.
19. Kauffman, S. A. *Origins of Order: Self-Organization and Selection in Evolution*. Oxford: Oxford University Press, 1989.

20. Kirkpatrick, S., C. D. Gelatt, Jr., and M. P. Vecchi. "Optimization by Simulated Annealing." *Science* **220** (1983):671–680.
21. Kirkpatrick, S. "Optimization by Simulated Annealing: Quantitative Studies." *J. Stat. Phys.* **34** (1984):975–986.
22. Krauth, W., and M. Mézard. "The Cavity Method and the Traveling-Salesman Problem." *Europhys. Lett.* **8** (1989):213–218.
23. Liao, W. "Replica Symmetric Solution of the Graph Bipartitioning Problem with Fixed, Finite Valence." *J. Phys. A* **21** (1988):427–440.
24. Lin, S., and B. W. Kernighan. "An Effective Heuristic Algorithm for the Traveling Salesman Problem." *Oper. Res.* **21** (1973):498–516.
25. Metropolis, N., A. W. Rosenbluth, M. N. Rosenbluth, A .H. Teller, and E. Teller. "Equation of State Calculations for Fast Computing Machines." *J. Chem. Phys.* **21** (1953):1087–1092.
26. Mézard, M., and G. Parisi. "A Replica Analysis of the Traveling Salesman Problem." *J. Physique (Paris)* **47** (1986):1285–1296.
27. Mézard, M., G. Parisi, and M. A. Virasoro. *Spin Glass Theory and Beyond.* Singapore: World Scientific, 1987.
28. Ogielski, A., and D. L. Stein. "Dynamics on Ultrametric Spaces." *Phys. Rev. Lett.* **55** (1985):1634–1637.
29. Palmer, R. G. "Broken Ergodicity." *Adv. Phys.* **31** (1982):669–735.
30. Palmer, R. G. "Statistical Mechanics Approaches to Complex Optimization Problems." In *The Economy as an Evolving Complex System*, edited by P. W. Anderson, K. J. Arrow, and D. Pines. Redwood City, CA: Addison-Wesley, 1988, 177–193.
31. Palmer, R. G., and D. L. Stein. "Glasses II: Models for Glassy Relaxation." In *Lectures in the Sciences of Complexity*, edited by D. L. Stein. Redwood City,CA: Addison Wesley, 1989, 771–785.
32. Palmer, R. G., D. L. Stein, E. Abrahams, and P. W. Anderson. "Models of Hierarchically Constrained Dynamics for Glassy Relaxation." *Phys. Rev. Lett.* **53** (1984):958–961.
33. Press, W. H., B. P. Flannery, S. A. Teukolsky, and W. T. Vetterling. *Numerical Recipes.* Cambridge: Cambridge University Press, 1986.
34. Sherrington, D., and S. Kirkpatrick. "Solvable Model of a Spin Glass." *Phys. Rev. Lett.* **32** (1975):1792–1796.
35. Siarry, P., and M. Dreyfus. "An Application of Physical Methods to the Computer Aided Design of Electronic Circuits." *J. Physique (Paris) Lettres* **45** (1984):L39–48.
36. Sibani, P., and K. H. Hoffmann. "Hierarchical Models for Aging and Relaxation of Spin Glasses." *Phys. Rev. Lett.* **63** (1989):2853–2856.
37. Stein, D. L. "Disordered Systems: Mostly Spin Glasses." In *Lectures in the Sciences of Complexity*, edited by D. L. Stein. Redwood City, CA: Addison Wesley, 1989, 301–353.
38. Stein, D. L., and R. G. Palmer. "Glasses I: Phenomenology." In *Lectures in the Sciences of Complexity*, edited by D. L. Stein. Redwood City, CA: Addison Wesley, 1989, 759–769.

C. Amitrano,† L. Peliti,‡ and M. Saber§

Dipartimento di Scienze Fisiche, Università di Napoli, Mostra d'Oltremare, Pad. 19, I-80125 NAPOLI (Italy). †Present address: The James Franck Institute, University of Chicago, 5640 Ellis Ave., Chicago, IL 60637. ‡GNSM-CISM, Unità di Napoli. Associato INFN, Sezione di Napoli. §Permanent address: Laboratoire de Magnétisme, Faculté des Sciences, Av. Ibn Battouta, B.P. 1014, Rabat (Morocco).

A Spin-Glass Model of Evolution

A simple model of molecular evolution, based upon spin-glass concepts, is introduced and its behavior is discussed. Adaptive and neutral evolutionary behaviors can be clearly identified. They are separated by a transition, which bears some relation with the gelation transition in spin glasses.

1. INTRODUCTION

A massive example of parallel computation goes under our own eyes, but is too slow to detect: evolution. Organisms attempt to find a path to better and better adaptation. They do so by producing slightly dissimilar offsprings. Fitter ones produce more offspring, unfit ones die. In this way, populations change with time, becoming in principle fitter and fitter to their environment.

Since the works of Sewall Wright,[14] it is costumary to represent the evolutionary process as a walk on an *adaptive landscape*: a population is represented by a point on a map, in which the horizontal coordinates represent the genotypes of the organisms, and the height represent their corresponding fitness. The rules of exploration are simple: from generation to generation random displacements in the horizontal direction are performed; if in one of these displacements one moves *uphill*, i.e., if a fitter variant is encountered, it becomes the starting point for the new generation; otherwise, the old genotype is retained. In this way equilibrium is reached when the point settles to a local maximum.

The procedure just outlined closely resembles Monte-Carlo optimization. One should however remark that it misses an important aspect. In fact, evolutionary adaptation is a *parallel* process: each organism of a population makes its trial at the same time.

This provides a way out of the paradox of complex adaptation. The space of *possible* genomes is immense: since the genome is encoded in the base sequence of DNA, and each base may be of one of four types (A, T, G, or C), the number of possible configuration for a genome of N bases is 4^N, which is much larger than any realistic population number even for very moderate values of N. Genomes coding for complex adaptation are presumably a tiny minority in this space. The probability that a single explorer hits on any of these is negligible.

Nevertheless, although populations are ridiculously small in number with respect to the whole genome space, they are sufficiently large to explore the surroundings of any given genome. For this it is only necessary that the population size, M, be much larger than N. But, on closer reflection, one realizes that it is also necessary that complex adaptation could arise out of a sequence of small improvements.

One is thus naturally led to ask the question: What properties of the fitness landscape are necessary to obtain significant adaptation?

One requires first of all that the fitness landscape contains a large number of optima, which correspond to different possible living structures. As emphasized by Anderson,[3] such a property is shared by the spin-glass Hamiltonians of the Sherrington-Kirkpatrick type (see, e.g., Mézard et. al.[12]). This justifies the use of spin-glass Hamiltonians as model fitness functions.

Kauffman and Levin[8] have considered adaptive walks on more general fitness landscapes, and have stressed the importance of the degree of correlation of the fitness function. In correlated landscapes similar genomes have similar values of the fitness function, whereas in uncorrelated (or *rugged*) ones even a single mutation leads to a completely independent value of the fitness. They have thus introduced a class of random functions of varying correlation, which is similar to a class of Hamiltonians for disordered systems first considered by Derrida.[5] In their work, the type of a whole population is represented by a single point: i.e., genome variability is neglected. They conclude that in highly rugged landscapes the population is likely to become trapped in low fitness optima, whose fitness values recede towards the mean fitness as the genome size N increases.

On the other hand, Eigen et al.[6] have considered deterministic equations governing the evolution of the genotype distribution in a population of self-replicating molecules. These equations comply with a statistical mechanical treatment similar to the transfer matrix formalism for lattice models.[11] The deterministic treatment is however only acceptable if the number of individuals of any relevant genome is sufficiently large. This is not usually the case in the midst of the evolution process, since interesting effects take place with the appearance of scarcely populated mutants, in particular when the population is trapped in low fitness optima.

Fontana and Schuster[7] have introduced a computer model of population evolution in a given fitness landscape, which models the dynamics of replication of RNA-like molecules in prebiotic conditions. They are able to highlight several interesting features of the evolution mechanism, but their conclusions are restricted by the choice of a particular fitness function.

We have chosen to consider a more general model, in which the fitness function may be varied within some limits. The interpretation of the replicating units is correspondingly looser. We report here the results of our investigations, most of which are contained in Amitrano et al.[1,2]

2. THE MODEL

The model we consider has been inspired by Anderson.[3] We have nevertheless strived to reach a maximum of simplicity, in order to highlight the effects of mutation and selection alone in a given fitness landscape.

First of all, we have chosen to work with a fixed population size M. This has mainly been motivated by computational simplicity in the simulations. Nevertheless we do not feel that it imposes too much of a restriction on the model. The effect we neglect is the possible variation in population size due to a variation of the average fitness.

Present-day information-carrying macromolecules are chains of monomers, each of which may be of one of four (for nucleic acids) or twenty (for proteins) types. But mutation and selection processes may already appear when the choice is restricted to just two types (standing, e.g., for purine or pyrimidine). Following the example of Fontana and Schuster,[7] we define therefore the configuration of a given individual α $(\alpha = 1, 2, \ldots, M)$ by a set s^α of N binary variables:

$$s^\alpha = (s_1^\alpha, \ldots, s_N^\alpha), \quad s_i^\alpha = \pm 1; \quad \alpha = 1, \ldots, M; \quad i = 1, \ldots, N. \quad (2.1)$$

We assume that the size N of the "genome" is fixed. This is a very drastic limitation, which we are planning to relax in future work. However, it allows us to consider a "genome space" defined as the space of all the 2^N possible genome configurations. This space has the topology of an N-dimensional hypercube. It is endowed with a natural notion of distance, i.e., the *Hamming distance* d_H, equal to the number of binary variables s_i which are different in the two configurations s, s':

$$d_H(s, s') = \frac{1}{2} \sum_{i=1}^{N} (1 - s_i s_i'). \quad (2.2)$$

Equivalently, one may consider the *overlap* $q(s, s')$ between the two configurations, defined by

$$q(s, s') = \frac{1}{N} \sum_{i=1}^{N} s_i s_i' = 1 - \frac{2d_H}{N} . \tag{2.3}$$

Selection involves the evaluation of a fitness function $H(s)$ defined on genome space. We defer to the next section the discussion of $H(s)$. We suppose only that an individual with higher values of H has a smaller probability of being removed at any selection step than those with lower values of H. We can therefore define a *death probability* $p(H)$ as a monotonically decreasing function of H, interpolating between 1 and 0 as H increases from $-\infty$ to $+\infty$. We have chosen a simple Fermi function:

$$p(H) = \frac{1}{1 + e^{\beta(H - H_0)}} . \tag{2.4}$$

The coefficient β is a *sharpness parameter*, and H_0 is a *threshold*. When $\beta \to \infty$, the death probability is 1 for all configurations s such that $H(s) < H_0$, and 0 otherwise: i.e., survival is sharply cut at the threshold. When $\beta < \infty$, the cut is more gentle, or, in other words, selection is sloppier. If $\beta = 0$, the death probability is always equal to $1/2$. Higher values of H_0 correspond to more exacting environments. Smaller values of β correspond to the existence of mechanisms which partially disconnect survival from fitness: e.g., strongly fluctuating environments.

We are thus led to the definition of the following mutation-selection mechanism in three steps:

1. *mutation*: a fraction τ of the $N \cdot M$ units (i, α) present in the population is chosen at random, and its state is changed:

$$s_i^{\alpha} \longrightarrow -s_i^{\alpha} ; \tag{2.5}$$

   we shall assume that τ is a small number; its actual magnitude will be discussed later;

2. *selection*: the fitness function $H(s)$ is evaluated for each individual α as a function of the configuration s^{α} of its genome; the individual is then removed from the population ("dies") with a probability p given by Eq. (2.4); and

3. *replication*: as a consequence of the previous step, a number M' of individuals will have died (we neglect the small probability that the whole population is annihilated in one go); to keep the population size constant, one chooses M' times an individual among the surviving ones and makes a copy of it.

The succession of the steps (1)–(3) is called a *generation*. Adaptation only affects the step (2). One might of course envisage more general scenarios, where not all mutations are equally likely, or where the replication probability is not equal for all surviving individuals. We think however that our model reduces arbitrariness to a bare minimum, still keeping sufficient generality to describe a set of reasonable evolution processes.

3. THE FITNESS FUNCTION

We now turn to a discussion of the fitness function $H(\mathbf{s})$. Quite independently of the nature of the individuals which make up our population—be they polymers in the prebiotic soup, or genes coding for homologous proteins in a given species—we expect that their fitness will be the outcome of a complex interaction with the environment. It is difficult to build up a satisfactory theory which associates an explicit fitness value to any given sequence $\mathbf{s}$.

We have considered a rather general class of functions, in order to analyze the effects of a varying degree of correlation of the fitness landscape. This landscape may be smooth or rough according to the higher or lower degree of correlation existing among the random values of $H(\mathbf{s})$ relative to different configurations. The landscape correlation measures how the fitness values of neighboring sequences $\mathbf{s}$ and $\mathbf{s}'$ are different from each other on average.

The smoothest landscape has only one optimum, and allows us to define at each configuration $\mathbf{s}$ a direction of steepest ascent, pointing towards that optimum. As the degree of correlation decreases, the number of optima increases, and the path length connecting any given sequence $\mathbf{s}$ to the nearest optimum decreases. On the other hand, in the same limit, the values of the fitness of these optima become smaller and smaller, and become typically of the order of the mean of $H(\mathbf{s})$. We may expect therefore that the behavior of evolving systems does depend on the correlation of the fitness landscape.

Kauffman and Levin[8,9] have introduced a class of fitness functions with tunable correlations in an evolutionary context. A basically equivalent class of functions had been introduced in the context of the theory of disordered systems by Derrida.[5] We have exploited their results to define a number of possible fitness landscapes, whose degree of correlation may be changed at will. These functions are analogous to the NK model introduced by Kauffman.[8,10] They depend on a parameter K, and yield completely uncorrelated fitness landscapes in the limit $K \rightarrow \infty$. This limit coincides with Derrida's Random Energy Model (REM).[5]

Although the REM is easiest to think upon, it is probably not suitable to describe actual evolutionary processes. Kauffman and Levin[8] argue that a certain degree of correlation is needed to avoid the *complexity catastrophe*, related to the fact that, in a rugged fitness landscape, the fitness values of the optima get closer and closer to the mean as the genome size, N, increases. We have therefore also considered a rather correlated fitness function, namely, the "spin-glass" function

$$H_2(\mathbf{s}) = \sum_{\{i,j\}} A_{ij}\, s_i s_j \,. \qquad (3.1)$$

The sum runs over all distinct pairs of units, and the A's are independent, identically distributed random variables for each such pair. Although it is rather correlated, it exhibits a relatively large number of local optima (at least if N is sufficiently large).

To summarize, we have considered the following two possibilities for $H(\mathbf{s})$:

1. the REM, completely uncorrelated, in which $H(\mathbf{s})$ assumes an independent value for each configuration $\mathbf{s}$: we take it to be uniformly distributed between -1 and 1;
2. the "spin-glass" function defined by eq. (3.1). We have taken the A's to be equal to $\pm a$ with equal probability, with $1/a^2 = N(N-1)/2$. In this way, the one-level probability distribution function $P(E)$ is approximately a Gaussian of zero mean and with variance equal to one.

4. THE $\beta = 0$ CASE

It is very instructive to start by investigating the simple case in which the sharpness parameter β appearing in Eq. (2.4) vanishes. The model then reduces to a special case of the diffusion-reproduction processes investigated by Zhang et al.[15]

If we let $\beta = 0$ in eq. (2.4), we obtain that the death probability p equals $1/2$ for each of the M individuals, independently of its configuration $\mathbf{s}$. Therefore, at each generation, roughly half of the individuals are removed ("die"). Since the remaining ones are reproduced to fill in the gaps in the population, we are in a situation in which, at each generation, any individual faces death or reproduction with equal probability. Moreover, we may conceive the effects of mutations as forcing the surviving individuals to perform a random walk on the hypercube of possible configurations.

Let us assume that the M individuals present at the beginning occupy random positions on the hypercube. As some of them die and others are reproduced to take their place, the population organizes into *families*, composed of the descendents of one of the original individuals.

In fact, after a number of generations of the order of M, all surviving individuals will belong to the same family, with a probability arbitrarily close to one. This result can be obtained in the easiest way by a simple mean-field argument.

If a population of variable size contains at time $t = 0$ exactly M individuals, and each of them has equal probability λdt of either dying or producing an offspring during the short time interval dt, then the probability that the population is extinct at time t is approximately given by

$$p_0(t) \simeq 1 - \frac{M}{\lambda t} , \qquad (4.1)$$

if λt is sufficiently large. Therefore, after a time t^* proportional to M/λ, $p_0(t)$ will exceed any given confidence threshold.

Let us go back to our original situation, with a fixed population size M and death probability $p = 1/2$. Let us assume that at any given time there are F families (i.e., groups of individuals having one of the original individuals as a common ancestor). On the average, each such family will have M/F members. Because of the above result, each such family will almost certainly disappear if we wait a number of generations proportional to the size of the family, i.e., to M/F. Therefore, over a time interval Δt, the number F of families decreases by ΔF, where

$$\Delta F = -F \cdot \left(\frac{F}{M}\right) \Delta t, \qquad (4.2)$$

yielding

$$F \sim \frac{M}{t}. \qquad (4.3)$$

These conclusions may be easily checked by computer simulations. We report in Figure 1 the results of a simulation, carried out by two of us in collaboration with Stamathis Nicolis at the University of Rome "La Sapienza."[13]

This *fluctuation instability* is in our opinion an essential feature of the class of models we are considering, in which the only interactions among individuals are either reproduction or competition for common resources. In fact, the only features needed to derive this result are a death probability which never vanishes, and a bounded population size (which makes the death and reproduction probability equal on average). However, the time needed to obtain a one-family population increases when the average death probability decreases.

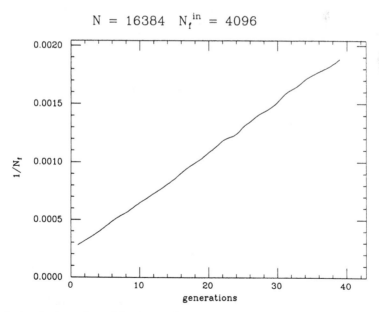

$$N = 16384 \quad N_f^{in} = 4096$$

FIGURE 1 Inverse number of families $1/F$ as a function of the number of generations t, for $M = 16,384$, initially divided in $F = 4096$ families.[13]

Let us now consider the population a very long time after the beginning. It is all made up of one family: indeed, by reckoning back of the order of M generations, we can identify the common ancestor of the whole population. Therefore each individual α will have accumulated at most $\tau N \cdot T$ mutations since it branched off the common ancestor. We assume that this number is much smaller than the size N of the genome. Given two individuals α, β, their mutual overlap $q^{\alpha\beta}$ is defined by

$$q^{\alpha\beta} = q(\mathbf{s}^\alpha, \mathbf{s}^\beta). \tag{4.4}$$

In the population we are discussing, this overlap is bounded by

$$1 \geq q^{\alpha\beta} \geq 1 - 2\tau T. \tag{4.5}$$

We may visualize the population as forming a small cloud in some region of genome space. Such a structure is called a *quasi-species*.[6] However, the region occupied by the population varies as time goes on. We may identify the center of the cloud (which we are tempted to call the *wild type*) by $\mathbf{s}^W = (s_i^W)$, where

$$s_i^W = \frac{1}{M} \sum_\alpha s_i^\alpha, \tag{4.6}$$

where the sum runs over all members of the population. The wild type mutates as fast as a single individual. Indeed, it is reasonable to expect that the wild-type configuration is close to that of the common ancestor, T generations back. When time t increases by Δt, the common ancestor will have undergone $\tau N \Delta t$ mutations. This will be the new location of the center of the population.

To summarize, when $\beta = 0$ the population organizes as a single quasi-species, with a comparatively small dispersion around a "wild type," which keeps on mutating at a frequency independent of the population size.

5. THE $\beta \rightarrow \infty$ CASE

The other simple limit is $\beta \rightarrow \infty$. As we have discussed in Section 2, this implies that all individuals whose fitness is smaller than the threshold die; the others survive and are reproduced. It is simplest to start by considering a rugged fitness function (i.e., the REM).

It is useful to visualize the genome configurations whose fitness is lower than H_0 as "forbidden" sites on the hypercube. They are randomly interspersed among the "allowed" sites, whose fitness exceeds H_0. The fraction x of allowed sites is a function of the threshold H_0:

$$x = \int_{H_0}^\infty dE \, P(E), \tag{5.1}$$

where $P(E)$ is the single-level distribution. Let us focus on one allowed site. Only a fraction x of its N neighbors (i.e., of the configurations which are a single mutation away from the site we are looking at) will be allowed on average. As a consequence, at any given mutation step, any individual will have a probability equal to $\tau N (1-x)$ to step on a forbidden site—and to be eliminated at the next step. This has the consequence that even fit individuals have a nonvanishing probability to die at each generation. This implies that the conclusions reached in the previous section also hold for this case. The population will therefore be made of a single quasi-species if one waits long enough.

The main difference with the previous case lies in the fact that the common ancestor of all surviving individuals at any given time must have wandered only on allowed sites. If the fraction x of allowed sites is close to 1, this will only slightly reduce the mutation speed of the wild type. However, as the environment becomes more exacting, and x decreases, the mutation speed will become smaller and smaller. It is possible to argue that, if the threshold H_0 exceeds a critical value H_c, this speed will essentially vanish in the long run. The critical value H_c is related to the critical value of x for percolation on the hypercube.

We can in fact define clusters of allowed sites, in the usual way of site percolation theory. Two allowed sites belong to the same cluster if they are nearest neighbors or if they may be connected by a path of allowed sites, each being the nearest neighbor of the following one. In the limit of very small mutation rate τ, the population always belongs to the same cluster of its common ancestor. The size of the clusters depends on x. There is a threshold value x^* of x, such that, if $x > x^*$, the largest cluster spans the whole hypercube.

Since the population is restricted to wander only on a connected cluster, the evolution of its genome configuration will bear traces of the extension of the cluster. A good measure of this property would be the self-overlap of the wild type with itself over a varying number of generations T:

$$Q(T) = \frac{1}{N} \sum_i s_i^W(t) s_i^W(t+T).$$

(5.2)

This quantity will in general fluctuate as t varies; however, if $x > x^*$, we shall have $Q(T) \to 0$ as $T \to \infty$: i.e., the population loses memory of its initial genome configuration. Otherwise, $Q(T)$ will vacillate around the average value of q, as the population wanders on the allowed cluster.

We have taken for granted in this argument that the quasi-species wanders on the largest cluster. This is warranted by the overwhelming extension of this cluster when $x > x^*$; on the other hand, if $x < x^*$, most sites belong to clusters with similar values of $\bar{q}$ and the result still holds with high probability.

It is interesting to remark that, at $\beta = \infty$, there is no direct "selective advantage" for higher values of the fitness, once they are above threshold. Therefore, all fit individuals are equivalent from the point of view of selection. Nevertheless, subpopulations living on different clusters are not equivalent: in fact, it is most likely that the one living on the largest cluster will finally take over. We see here, in a nutshell, a level of "species selection," which cannot be reduced to a lower level.

When β is not strictly infinite, there is a small selective advantage of fitter configurations, due to the exponential "tail" in the death probability. A similar effect holds for correlated fitness landscapes, since it is less likely that a configuration with a higher fitness value will be surrounded by forbidden configurations. This entails a weak trend towards adaptation.

Adapting mechanisms also emerge at $\beta = \infty$ in correlated landscapes. In this situation, configurations which have high fitness values are more likely surrounded by other fit configurations: therefore, the effective death probability, which depends on the probability that a mutation leads to an unfit configuration, is correspondingly smaller. This effect is stronger when the fitness landscape is more correlated.

6. DISCUSSION

The main conclusions reached in the present work are the following:

1. The fluctuation instability prevents the eventual stabilization of coexisting, well-defined, molecular quasi-species in all models in which the interaction between different individuals is reduced to the competition for common resources; it follows that one should introduce from scratch some interaction mechanisms when defining chemical evolution models.
2. Models of this kind may exhibit either neutralist or adaptive behaviors depending on the nature of the fitness landscape and on the values of the relevant parameters.
3. It is important to probe the correlation of the fitness landscape in which a given system is evolving. Complexity of the interaction between individual and environment corresponds to less correlated landscapes.

One might consider the possibility of probing landscapes with different correlations by varying the selection mechanisms in experiments of evolution *in vitro* like those of Biebricher et al.[4] Replication speed in $Q\beta$-polymerase-$Q\beta$-RNA systems probably corresponds to a highly correlated landscape, in which only the detailed RNA configuration near the recognition site (and the chain length) is probed. Resistence to replication inhibition should make the landscape more corrugated, by increasing the number of selected traits.

We hope that our investigation will encourage some researchers to pay increased attention to analogous abstract evolution models, which may be useful to define more sharply the questions that evolutionary theory is required to answer.

ACKNOWLEDGMENTS

The work of M.Saber was performed within the "I.C.T.P. Programme for Research and Training in Italian Laboratories." L. Peliti and M. Saber warmly thank S. Nicolis and M. Serva for fruitful discussions.

REFERENCES

1. Amitrano, C., L. Peliti, and M. Saber. "Neutralism and Adaptation in a Simple Model of Molecular Evolution." *C.R. Acad. Sci. Paris, III* **307** (1988):803–806.
2. Amitrano, C., L. Peliti, and M. Saber. "Population Dynamics in a Spin-Glass Model of Chemical Evolution." *J. Mol. Evol.*, in press.
3. Anderson, P. W. "Suggested Model of Prebiotic Evolution: The Use of Chaos." *Proc. Natl. Acad. Sci. USA* **80** (1983):3386–3390.
4. Biebricher, C. "Darwinian Evolution of Self-Replicating RNA." *Chemica Scripta* **26B** (1986):51–57.
5. Derrida, B. "Random Energy Model: Limit of a Family of Disordered Systems." *Phys. Rev. Lett.* **45** (1980):79–82.
6. Eigen, M., J. McCaskill, and P. Schuster. "Molecular Quasi-Species." *J. Phys. Chem.* **92** (1988):6881–6891.
7. Fontana, W., and P. Schuster. "A Computer Model of Evolutionary Optimization." *Biophys. Chem.* **26** (1987):123–147.
8. Kauffman, S. A., and S. Levin. "Towards a General Theory of Adaptive Walks on Rugged Lanscapes." *J. Theor. Biol.* **128** (1987):11–45.
9. Kauffman, S. A., E. D. Weinberger, and A. S. Perelson. "Maturation of the Immune Response via Adaptive Walks on Affinity Landscapes." *Theoretical Immunology Part One*. SFI Studies in the Sciences of Complexity, edited by A. S. Perelson. Redwood City, CA: Addison-Wesley, 1988, 349–382.
10. Kauffman, S. A. "Adaptation on Rugged Fitness Landscapes." In *Lectures in the Sciences of Complexity*. SFI Studies in the Science of Complexity, Lect. Vol. I, edited by D. Stein. Redwood City, CA: Addison-Wesley, 1989, 512–617.
11. Leuthäusser, I. "Statistical Mechanics of Eigen's Evolution Model." *J. Stat. Phys.* **48** (1987):343–360 .
12. Mézard, M., G. Parisi, and M. A. Virasoro, *Spin Glass Theory and Beyond*. Singapore: World Scientific, 1987.
13. Nicolis, S., L. Peliti, and M. Saber. "Evolution in a 'Flat' Fitness Landscape." Unpublished.

14. Wright, S. "The Roles of Mutation, Inbreeding, Crossbreeding and Selection in Evolution." *Proceedings of the Sixth International Congress of Genetics* **1** (1932):356–366.

15. Zhang, Y. C., M. Serva, and M. Polikarpov. "Diffusion-Reproduction Processes." Unpublished.

D. L. Stein
Department of Physics, University of Arizona, Tucson, AZ 85721

What Can Physics Do For Biology?

INTRODUCTION

A physicist attempting to study biological problems is engaged in a very risky business. The reason usually put forward to explain this is that a physicist looks for universality and simplicity, which more often than not serve as useful guides for uncovering physical law. On the other hand, biology results from chance and history, so that differentiation and detail become the guiding principles of investigation. The use of Occam's razor in biological investigations often leads to self-inflicted wounds.

It then comes as no surprise that, more often than not, new physical insights have been uncovered by biological investigations than the other way around. Probably the best known example of this is the case of Brownian motion, which was first discovered as a result of biological investigations, and whose statistical analysis by Einstein provided important support for the atomic theory of matter. (In the sense in which statistical analyses have become a central feature of modern science, all of the sciences owe a great debt to the practice of gambling throughout the ages, but that belongs to a different discussion.) This seems to be part of a larger trend in the history of science, in which the flow of ideas often travels in the direction opposite to the reductionistic path. Without attempting to explore further the reasons for these trends (or even bothering to justify their truth), I simply point out that it was this kind of observation which led Stanislaw Ulam to utter the remark, "Ask not what [physics] can do for [biology]—ask what [biology] can do for [physics]."[14]

It would nevertheless be overly pessimistic, if not cowardly, to abandon all efforts to apply physical insights to biological problems. Aside from the obvious fact that atoms and molecules must still obey physical laws whether within biological systems or without, mathematical models of interacting biological components, whether atoms in a protein, neurons in a brain, or species in an ecosystem, can provide useful qualitative insights as long as the questions asked and conclusions drawn are carefully circumscribed. Physicists and mathematicians are not about to provide a comprehensive picture of any of these problems anytime soon, but may still shed some light on certain of their aspects through inspired modeling.

There is yet another set of questions, which physical scientists and mathematicians may be best poised to ask (and in the pursuit of which the organizers of this meeting have been in the forefront): given a general process, such as adaptation, are there any general principles, rules, or universal features which may be subject to mathematical analysis? Here one is examining a *process* which may be subject to its own internal laws; in the case of adaptation, for example, similar rules at some very basic level may govern the evolution of macromolecules in the prebiotic soup, of species in an extant ecosystem, or of antibodies in the immune response. A successful picture of such a process should be consistent with the historical path followed by any particular realization of the process, but independent of any such outcome. In other words, if we start over again with a prebiotic earth (or somewhere else for that matter), the course of evolution will certainly be different, but the general rules followed would be the same. Obviously, it's crucial in this game to ask precisely defined questions and to state clearly the boundaries within which the inquiry falls.

AN EXAMPLE

Consider the following set of problems:

1. Dynamics of Glasses;
2. Chemical Kinetics in Globular Proteins;
3. Evolutionary Processes;
4. Protein Folding; and
5. Maturation of the Immune Response.

A satisfactory understanding of any one of these will ultimately depend on detailed study of a host of features unique to it. Nevertheless, it might be worthwhile to ask whether any subset of these problems share any common features which might give rise to relationships among them which are both sensible and interesting to explore. These particular problems are relevant to the context of this workshop since all, in one way or another, involve processes which can be modeled as a walk (random or otherwise) on a rugged landscape. The landscape in the case of glasses might be energy as a function of atomic configuration, and the walk would then represent rapid transitions among such microstates. In the case of evolution, the landscape could represent survival probability as a function of species, given a certain environment; the walk in this context represents transitions among "microstates" which enumerate different populations of existing species (including their appearance and disappearance). Since others will be discussing the problems associated with walks on rugged landscapes I won't say too much more here, except to note that any deeper relation among two different problems connected in this way depends on whether such walks follow any general rules. That is, do any "universality classes" of walks on rugged landscapes exist, such that a wide variety of landscapes exhibit similar behavior independent of details? If so, then maturation of the immune response, say, may tell us something interesting about evolution. If not, then connections among these problems don't extend very deeply, although study of random walks on the "correct" landscape (if it can be found!) would still provide useful information an any one particular problem. The NK model introduced by Stuart Kauffman and his collaborators may provide useful insights into some of these questions.

What else do these problems have in common? My own list includes:

1. All contain many strongly interacting units: atoms in a glass or protein, species in the case of evolution, and so on. There also exist external factors, which may or may not change with time, which influence the outcome of a particular process.
2. Each problem contains a wide range of intrinsic timescales: individual atoms in a glassy liquid may relax relatively quickly, while others may move only in a group and therefore will relax more slowly. Still other relatively small groups may not be able to relax until yet others get out of their way, and so on. Within an evolutionary context, some species may be very short lived and others very long lived, with a wide continuum in between.
3. The dynamics in all cases is governed by barrier crossing. This is of central importance, since we're dealing with *meta*stable states in all cases; i.e., states which are long lived (or at least don't spontaneously decay away) but don't last forever. A living system, for example, is metastable; a small perturbation won't kill you, but a large one will. The meanings of "small" and "large" obviously depend on the system, but in all cases there is some threshold which divides the two. In the context of walks on a rugged landscape, this means that in order to change from one metastable configuration to another, the system has to pass through a sequence of less favorable (and therefore less probable) configurations; this is what we mean by a "barrier." The system may be driven to surmount the barrier by external forces, or the process may simply be a result of equilibrium

fluctuations. Typically, the system will spend very little time in these low-probability configurations, which is the reason why you observe reactants and products in a chemical reaction but not the intermediate transition states. It's also a way of understanding, at some level, why equilibria in evolution seem to be "punctuated," and why "missing links" are hard to find.[13] This also illustrates one of the points about universality made in the introduction; if evolutionary processes are indeed governed by these rules, then punctuated equilibria should be the norm in evolution, here or in the next planet over.

4. Finally, the barriers are governed by the individual units themselves, and are therefore not static; a barrier may be changing on the order of an average crossing time. The environment a given species finds itself in depends strongly on other species in the same habitat; these are continually changing. In a glassy liquid, certain atoms may block the diffusion of other atoms; the former constitute the barriers. Since atoms are subject to thermal noise, it may be a poor approximation to model relaxation or diffusion as a crossing of a static barrier.

In the simplest case, consider a potential as shown in Figure 1, and think of whatever system we're interested in as a point particle. The system lies in a metastable state and is connected to a heat bath at fixed temperature T which is small compared to ΔF. The particle will typically perform small oscillations of frequency ω_0 about the bottom of its local minimum; ω_0 is related to the curvature at the bottom of the well. The particle will occasionally receive a large thermal "kick", sending it part way up the well, at which point it will "roll" back down to its metastable equilibrium state. The probability of a kick large enough to send it over the barrier is proportional to $\exp(-\Delta F/T)$, so crossing events are rare and therefore statistically independent. This simple problem has two well-separated timescales: a dynamical timescale ω_0^{-1} and a statistical timescale proportional to $\exp(\Delta F/T)$, which is called the Arrhenius law. Calculation of prefactors requires a careful treatment; the pioneering analysis of Kramers[5] serves as the basis for most such studies.

Now one of the characteristic features of a viscous liquid nearing the glass transition is that its viscosity, over several decades, does *not* exhibit Arrhenius behavior, nor do other relaxational timescales related to configurational rearrangement. More precisely, the slope of the logarithm of the viscosity vs. $1/T$ increases as T decreases. (As with any other blanket statement applied to glasses, there are whole classes of exceptions to this rule. Nevertheless, most glasses do indeed display non-Arrhenius behavior, and uncovering the reason for this has become one of the central problems in glass science.) One possible path toward understanding this may lie in realizing that, as pointed out in 4. above, a glassy liquid consists of many degrees of freedom possessing a wide spectrum of relaxational (or diffusional, or whatever process is relevant) timescales, and that therefore a barrier to relaxation or diffusion may depend on other degrees of freedom which are changing on timescales of the same order as a typical escape time.[11,12]

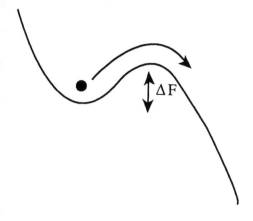

FIGURE 1   Traditional picture of activation; the particle waits for a sufficiently large "kick" (energy fluctuation due to thermal noise) to enable it to surmount the barrier.

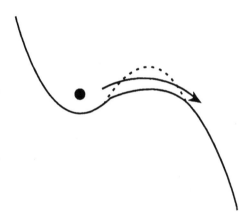

FIGURE 2   If the potential is fluctuating, the crossing rate may be increased if the waiting time for a sufficiently large downward fluctuation of the barrier is comparable to or less than the waiting time for a "kick" sufficient to cross the higher average barrier.

A vivid example of this behavior can be seen in a movie based on a molecular dynamics simulation, by Cheeseman and Angell,[1,2] of the disrupted-tetrahedral silicate $Na_2O \cdot 3SiO_2$ over the timescale range 10^{-15} to 10^{-11} sec. One notices from the simulation that the relatively small sodium ion is normally unable to diffuse because it is trapped in a small region controlled by a "gate" of larger oxygen ions. Occasionally, thermal fluctuations result in the oxygen ions moving relatively far apart for a short period of time, enabling the sodium ion to diffuse through. A schematic picture of the process involved is given in Figure 2.

We have found, after studying a very simple model of escape over a fluctuating potential barrier, that the escape time τ depends sensitively on the correlation time of the potential barrier fluctuations, denoted by τ_c. (Roughly, τ_c gives a measure of how rapid the barrier fluctuations are.) We examined three regimes[11]: $\tau_c \gg \tau_A$, where τ_A is the escape time over the (static) average barrier; $\tau_c \sim \tau_A$; and $\tau_c \to 0$

(the "white noise" limit). We also examined corrections to the white noise limit for small but nonzero τ_c.[12] In all cases studied so far we found that the effective barrier, defined as $\Delta F_{eff} = \log \tau/(1/T)$, decreases as τ_c decreases. Now whatever model one might make of this situation will almost certainly require τ_c to decrease as temperature increases; greater temperatures tend to speed up processes. This then indicates, at least qualitatively, non-Arrhenius curvature of τ in the correct direction.

While this work represents only a first step, it does indicate that such fluctuations should not be ignored. This is well known to be the case in certain problems in biology as well; for example, conformational fluctuations of large biomolecules are crucial for proper functioning of the molecule. Such fluctuations are necessary, for instance, for oxygen to diffuse through hemoglobin or myoglobin to reach the heme group.[3,4] Techniques which we are developing to study a particular problem in glass physics can therefore be used to gain some insight into this problem as well. Other possible applications include the study of models of evolution in which the fitness landscape, described earlier, continually changes due to the introduction and disappearance of different species.[6,7]

PROTEIN FOLDING

The moral so far is that landscapes change as interesting processes occur; the meanings of "landscape" and "interesting" depend on the problem at hand. While the idea of rugged landscapes has wide applicability, it may also be generally true that a random walk on a *static* landscape may not be sufficient.

One question which is then worth asking is: Do proteins make use of this fact during the folding process? One of the most perplexing problems in molecular biophysics arises from the speed with which a protein finds its way, through a staggering maze of possibilities, to its native state. A vast compendium of knowledge and folklore concerning this problem has accumulated over years of study, and some of it has been presented at this meeting by the Richardsons. We will not attempt to tackle this problem directly, since a real understanding must rely on detailed biochemistry and chemical kinetics, but instead will pose a much simpler question, though clearly inspired by the folding problem. Our hope, of course, is that in attempting to answer the simpler question we may learn something fairly basic about the much harder problem.

Let us then consider the following rather general problem: Consider a system with a very rugged landscape, having many peaks and valleys. Suppose that the optima are uncorrelated or, at best, weakly correlated, and furthermore that most have only relatively mediocre values. Stuart Kauffman refers to this as the "complexity catastrophe"; such a situation occurs in the random energy model of

Derrida,[10] which is a thermodynamic system with a Gaussian distribution of uncorrelated energies. Can one find a relatively simple procedure for finding a (very rare) optimum with an extremely good value in a fairly short time?

Clearly, simply starting at an arbitrary spot on the landscape and executing a random walk with the expectation of finding a good optimum through luck is completely hopeless for a large number of optima. Because of the lack of correlation of goodness values of the optima, there is no apparent way even to tell if one is getting "warm." Hence, a walk on a static landscape of this type has little chance of finding a good optimum.

But why should the landscape be fixed? We[9] have been attempting to construct an algorithm which we believe has a chance of at least reaching a part of configuration space which contains very good optima, and therefore vastly narrowing down the number of states which need to be examined.

We chose to study the NK model (see Kauffman and Weinberger, this volume)[8] which for $k = N-1$ satisfies the conditions described above for the complexity catastrophe. We start with $k = 2$, where the optima are strongly correlated, and simple simulated annealing, for example, will enable one to reach a very good optimum with high probability in a relatively short time. The basic idea now is to use k as a tuning parameter. If k is increased by one, the state the system is sitting in is no longer a local optimum, so one first lets the system relax to the nearest local optimum. One then executes a Monte Carlo algorithm to search for better states in the neighboring region of configuration space. As k is slowly increased in this manner, information from the previous step is utilized to keep the system in a region of relatively low energy. The hope is that when $k = N - 1$ is reached, the system resides in a state of approximately the same energy as it started in with $k = 2$, which would actually be an extremely good optimum. Work on this algorithm is still in progress.

The examples which I've discussed fall outside of the scope of traditional biophysics. The problems which are being tackled do not rely on a direct application of physical principles to biological processes, or even on model building, but rather on the elucidation of mathematical principles that may lie behind whole categories of processes. It's probably inaccurate to call such investigations "biology" at this stage, unless one wishes to study "organism-free" biology, which in at least some of these cases would be a fair accusation. On the other hand, investigations aimed at uncovering general rules or principles lying behind adaptation, evolution, searching through fitness landscapes, and the like, may ultimately provide useful insights that extend across a spectrum of disciplines.

uncovering general rules or principles lying behind adaptation, evolution, searching through fitness landscapes, and the like, may ultimately provide useful insights that extend across a spectrum of disciplines.

REFERENCES

1. Angell, C. A., and P. A. Cheeseman. Movie shown at Institute for Theoretical Physics, Santa Barbara, CA, 1987.
2. Angell, C. A., C. A. Scamehorn, C. C. Phifer, R. R. Kadiyala, and P. A. Cheeseman. *J. Phys. Chem. Minerals* (1988):15, 221.
3. Beece, D., L. Eisenstein, H. Frauenfelder, D. Good, M. C. Marden, L. Reinisch, A. H. Reynolds, L. B. Sorensen, and K. T. Yue. "Solvent Viscosity and Protein Dynamics." *J. Biochem.* (1980):19, 5147.
4. Karplus, M., and J. A. McCammon. *J. CRC Crit. Rev. Biochem.* (1981):9, 293.
5. Kramers, H. A. *Journal Physica* (1940):7, 284.
6. Kauffman., S. A. *Origins of Order.* England: Oxford University Press, 1990, in press.
7. Kauffman., S. A. "Adaptation on Rugged Fitness Landscapes." In *Lectures in the Sciences of Complexity*, edited by D.L. Stein. SFI Studies in the Sciences of Complexity, Lect. Vol. I. Redwood City, CA: Addison-Wesley, 1989.
8. Kauffman, S. A., and E. D. Weinberger. This volume.
9. Clark, L., S. A. Kauffman, M. Pokorny, and D. L. Stein. Work in progress.
10. Derrida, B. *Phys. Rep.* **67**, 1, 80.
11. Stein, D. L., R. G. Palmer, J. L. van Hemmen, and C. R. Doering. "Mean Exit Times Over Fluctuating Barriers." *Phys. Lett. A* (1989):136, 353.
12. Stein, D. L., C. R. Doering, R. G. Palmer, J. L. van Hemmen, and R. McLaughlin. "Escape Over a Fluctuating Barrier: The White Noise Limit." *J. Phys. A.* **23** (1990):L203.
13. Wolynes, P. G. "Chemical Reaction Dynamics in Complex Molecular Systems." In *Lectures in the Sciences of Complexity*, edited by D. L. Stein. SFI Studies in the Sciences of Complexity, Lect. Vol. I. Redwood City, CA: Addison-Wesley, 1989.
14. The words "physics" and "biology" appear in brackets because Ulam did not, to my recollection, use them. I have been unable to locate the original quote, but the spirit is the same as that used in the text; namely, that if subject A lies, in some reductionistic sense, at the foundation of subject B, then investigations begun within subject B are more likely to influence subject A than the other way around.

P. Schuster
Institut für Theoretische Chemie der Universität Wien, Währingerstraße 17, A-1090 Wien, Austria

Optimization of RNA Structure and Properties

Folding of RNA molecules and evolution of their structures according to predetermined rules result in highly *rugged* value landscapes which originate from the complexity of genotype-phenotype relations. A model of phenotype evaluation through mutation and selection is introduced. Extensive computer simulations were performed in order to test the predictions of the model. In addition they provide insight into the mechanism of escape from traps of the evolutionary optimization process.

RUGGED COST FUNCTIONS AND VALUE LANDSCAPES

During the last decade more and more problems from mathematics, physics, biology and other scientific disciplines were found to be especially hard and difficult to solve even with the assistance of powerful computers. Moreover, these problems have the naughty property that the numerical efforts to find a solution increase very fast, presumably exponentially with system size. According to this and other features which are used for classification[14,13] and which shall not be discussed here, these problems are characterized as N(ondeterministic) P(olynomial time)-complete. Problem solving techniques from physics and biology were generalized and applied to optimization tasks in other disciplines.[3] Three classes of methods became particularly important:

- *Simulated Annealing.* This method originated in spin glass research[23] and represents a modification of *Monte-Carlo* optimization methods.[27]
- *Neural Networks.* They were originally conceived as models for the dynamics of central nervous systems. Recently, they were *rediscovered* and are used now in some optimization problems. The main field of application, however, is pattern recognition.[19]
- *Genetic Algorithms.* These methods mimic biological evolution and apply its principles to optimization problems in other disciplines.[16,18,29]

In this contribution we shall apply simulated annealing and a genetic algorithm which is derived from the evolution of simple organisms—so-called *prokaryonts*: viruses, bacteria and cyanobacteria—to problems of molecular evolution and compare the results of both techniques.

Most complex optimization problems share a common feature: one searches for the global maximum—or global minimum—of a highly structured objective function or cost function $V(x_1, x_2, \ldots, x_n)$ of many variables. The number n may be as large as several thousand and more. In some cases the qualitative behavior in the limit $\lim n \to \infty$ is of interest. Using vector notation, $\mathbf{x} = (x_1, x_2, \ldots, x_n)$, we can define the problem to find the maximum of the cost function by

$$\mathbf{x}_m : \qquad V(\mathbf{x}_m) = \max\{V(\mathbf{x})\} . \qquad (1)$$

A typical cost function of a complex optimization problem has many maxima widely distributed with respect to height. Only a small percentage of these peaks is of interest. Most of them lie at small values of the cost function and are useless as approximate solutions of the optimization problem.

Evolutionary adaptation in biology can be understood as a complex optimization problem.[21] The object to be varied is a polynucleotide molecule—DNA or, in some classes of viruses, RNA. It is commonly called the *genotype*. The genotype is not evaluated as such; it is first mapped into a *phenotype* and then the selection principle operates on phenotypes. Charles Darwin's great merit was to find the universal mechanism of phenotype evaluation in nature: *natural selection* implies that those variants which have more—and/or more fertile—descendants in future generations increase in number whereas their less-efficient competitors decrease. Efficiency in the natural evaluation process refers exclusively to reproductive success. The variant which produces the largest number of fertile descendants replaces ultimately all competitors and is selected. In population genetics the notion of *fitness* was introduced as a quantitative measure for the capability of being selected. The cost function of evolutionary optimization is therefore often characterized as *fitness landscape*. We shall show, however, that fitness in the sense of Darwin's *survival of the fittest* cannot be assigned to a single genotype. It is an ensemble property. Therefore we suggest to use the notion of *value landscape* rather than fitness landscape as an appropriate synonym for cost function in biology.

It is impossible to derive analytic expressions for realistic value landscapes in biology. This holds true even for the most simple conceivable cases like folding and replication of nucleic-acid molecules. We shall study a model of this process here. In this model secondary structures are computed from known genotypes by means of a folding algorithm.

Genetic algorithms mimic evolutionary optimization. A *population* of test solutions is set up, individual solutions are *replicated* and subject to variations. The better a given solution, the more descendants it has. Two types of variations are commonly considered: mutations—variations within a solution—and recombinations—exchange of parts between different solutions. Selection is achieved by constant population size. Excess production is compensated by an appropriate dilution flux.

GENOTYPE AND PHENOTYPE

In view of their chemical structure nucleic acids are visualized best as *strings* or sequences of symbols. There are four classes of symbols. In ribonucleic acid RNA they consist of **G**(uanine), **A**(denine), **C**(ytosine) and **U**(racil). Double helical structures (Figure 1) are particularly stable in themodynamical terms. The symbols are grouped into two base pairs which constitute complementarity relations: $\mathbf{G}{\equiv}\mathbf{C}$ and $\mathbf{A}{=}\mathbf{U}$. All other combinations of symbols opposing each other in the double helix cause low stability or instability.

In order to illustrate the source of ruggedness in the value landscapes of RNA folding and replication, we distinguish the genotypic sequences or strings $\mathbf{I}_k$ and the phenotypic[1] spatial structures $\mathcal{G}_k(\mathbf{I}_k)$. The phenotype is formed through folding to a three-dimensional structure as shown in Figure 1. The driving force for this spontaneous process in aqueous solution of proper ionic strength is minimization of Gibbs' free energy through base pairing, stacking and other types of molecular association. Polyribonucleotide folding can be partitioned into two steps:

- folding of the string into a quasiplanar—two-dimensional—secondary structure through formation for as many base pairs as possible and
- formation of a spatial—three-dimensional—structure from the planar folding pattern.

The first of these two steps is modeled much more easily than the second one. Efficient algorithms are available for the computation of RNA secondary structures.[34] The prediction of tertiary structures of RNA molecules is much harder. At present there are no reliable theoretical models available for this purpose.

[1]Here we generalize the notion of phenotype commonly used for organisms to molecules.

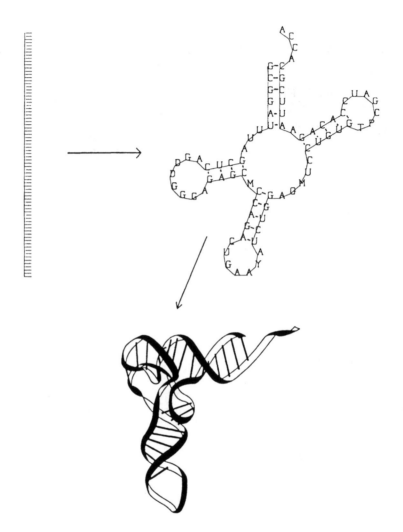

FIGURE 1   Folding of a polyribonucleotide chain into the two-dimensional secondary and the three-dimensional tertiary structure of the RNA molecule. *Phenylalanyl-transfer*-RNA, an RNA with a chain length of $\nu = 76$ bases, is chosen as an example. The symbols **D, Y, M, T** and **P** are used for modified bases which are unable to form base pairs. The major driving force of folding into secondary structures is the formation of thermodynamically favored, double helical substructures which contain stacks of complementary base pairs: $G\equiv C$ and $A=U$. It is important to note that in double helices the two nucleotide strands run in opposite direction.

EVALUATION OF PHENOTYPES

The mean number of descendants is the evaluation criterion of Darwin's principle of selection in stationary populations: variants with more descendants than the mean increase in percentage of total population. Less successful competitors decrease in number until they die out. Elimination of less-efficient variants causes the mean to increase. Then more and more variants fall under the failure criterion and ultimately only a single fittest type remains as expressed by the common—and often misused— catchphrase of *survival of the fittest.* In order to make the model quantitative and to develop it further, a measure of *fitness* is required. This measure has to be defined independently of the outcome of selection in order to avoid the long-known vicious circle of the *survival of the survivor.* In principle, chemical-reaction kinetics is in a position to provide an independent access to the selection criterion if one would apply it to the biological multiplication process.

The main difficulty of the experimental investigation of Darwin's principle is caused by the enormous complexity of all organisms—even including bacteria— which is presently prohibitive for a study of multiplication by the methods of physics and chemistry. The simplest system in which selection could be observed consists of RNA molecules which are multiplied *in vitro* by a replication essay.[31] This essay contains an RNA replicase which replicates RNA of the bacteriophage $Q\beta$ in cells of the bacterium *Escherichia coli,* together with an excess of all molecular materials necessary for the synthesis of RNA. By means of the $Q\beta$ system the kinetics of RNA replication was explored in great detail.[2] These investigations can be understood as the basis of a molecular model of evolution[5,6] which we shall briefly discuss now.

Darwin's principle can be formulated on the molecular level by means of kinetic—ordinary—differential equations. The variables are the concentrations of genotypes: $[\mathbf{I}_k] = c_k(t)$. The appropriate measure of genotype frequencies in populations are relative concentrations normalized to 1:

$$x_k(t) = \frac{c_k(t)}{\sum_{i=1}^n c_i(t)} \text{ with } \sum_{i=1}^n x_i(t) = 1. \tag{2}$$

Two classes of chemical reactions determine the time dependence of relative concentrations: replication with rate constant A_k

$$(\mathbf{A}) + \mathbf{I}_k \xrightarrow{k} 2\mathbf{I}_k \ ,$$

and degradation with rate constant D_k

$$\mathbf{I}_k \xrightarrow{D_k} (\mathbf{B}); \quad k = 1, 2, \ldots, n \ .$$

The symbols $\mathbf{A}$ and $\mathbf{B}$ denote low-molecular-weight materials which are either required for RNA synthesis or which are produced by degradation, respectively. The materials for the synthesis are assumed to be present in excess and their concentrations are essentially constant. Concentrations of degradation products do not enter the kinetic equations since the process is irreversible under the conditions of the experiment. Hence, neither the concentrations of **A** nor those of **B** represent variables in the kinetic equations—and both symbols were put in parantheses accordingly.

The kinetic differential equations depend only on differences in the rates of synthesis and degradation. It is useful therefore to introduce an *excess production*, $E_k = A_k - D_k$, of the individual genotypes I_k:

$$\frac{dx_k}{dt} = (E_k - \overline{E})x_k \text{ with } \overline{E}(t) = \sum_{i=1}^{n} x_i E_i; \quad k = 1, \ldots, n. \tag{3}$$

The mean excess production of the entire population is denoted here by $\overline{E}(t)$. It represents the molecular analogue of the mean number of descendants mentioned initially as the evaluation criterion for phenotypes.

The solutions of the differential Eq. (3) are given by

$$x_k(t) = \frac{x_k(0)\exp(E_k t)}{\sum_{i=1}^{n} x_i(0)\exp(E_i t)}; \quad k = 1, \ldots, n. \tag{4}$$

Provided that we waited for a sufficiently long time all variables except one would have approached zero: $\lim_{t \to \infty} x_j(t) = 0$ for all $j \neq m$. The only non-vanishing variable, $\lim_{t \to \infty} x_m(t) = 1$, describes the concentration of that variant which has the maximum excess production, $E_m = \max\{E_i; i = 1, \ldots, n\}$, and hence reproduces itself most efficiently. Ultimately only the fittest genotype I_m remains.

The Darwinian scenario as well as conventional population genetics are built upon the assumption that mutations are rare events. Selection is fast—occurring on a time scale which is set by the rate constant of replication, or multiplication in general—and evolutionary adaptation is slow, since it is determined by the rate at which advantageous mutants are formed. Only then, we can expect the concept of *survival of the fittest* to reflect reality. Evolutionary optimization is characterized as a sequence of genotypes which were fittest at the instant of their selection.

In the light of the knowledge provided by present-day molecular biology the simple evaluation scheme has to be modified in two points:

- The molecular biology of viruses—and to some extent also that of bacteria—has shown that mutations occur much more frequently than the classical scenario assumes. In the case of frequent mutations no single fittest type survives. A distribution of genotypes generally consisting of a most frequent sequence, the *master sequence*, and its closely related mutants is selected instead. This distribution was called *quasispecies*[6] in order to point at some analogy to the notion of species with higher organisms.

- A high precentage of mutants is selectively neutral. Their proliferation in natural populations is the major source of information for phylogenetic studies on the molecular level. The dynamics of neutral selection is a stochastic phenomenon. It can be predicted by probability calculus only. Stochasticity is of general importance when particle numbers are small.

Proliferation of neutral mutants is generally observable in finite populations only. It can be modeled by means of stochastic processes but not by kinetic differential equations. Every model which uses kinetic differential equations is valid in the limit of large—in principle infinitely large—particle numbers. For most chemical reactions this is no real restriction, since all relevant results refer to 10^{10} or more molecules. In the domain of molecular evolution the limit of large numbers is not generally fulfilled. Every new genotype produced by mutation is at first present in a single copy only and the initial phase of growth is also dominated by stochastic phenomena.

The first problem has been extensively dealt with in the theory of the molecular quasispecies.[7,8,9] We restrict this contribution to a brief account on the most important results. The second question cannot be treated here in sufficient detail. A review of the theory of neutral mutants is found in M. Kimura.[22] The computer simulations which will be introduced later comprise also the stochastic effects of evolution.

In the frequent-mutation scenario, mutant production must be accounted for explicitly. Using Q_{ki} for the frequency at which the genotype $\mathbf{I}_k$ is synthesized as an error copy of the sequence $\mathbf{I}_i$ the reactions leading to mutants can be written as

$$(\mathbf{A}) + \mathbf{I}_i \xrightarrow{A_i Q_{ki}} \mathbf{I}_k + \mathbf{I}_i \,.$$

Adding these reactions to error-free replication and degradation we end up at the kinetic differential equations of the replication-mutation system:

$$\frac{dx_k}{dt} = \sum_{i=1}^{n} A_i Q_{ki} x_i - \left(D_k + \overline{E}(t) \right) x_k \,; \quad k = 1, \dots, n \,. \tag{5}$$

We note that the mean excess production $\overline{E}(t)$ is defined precisely as in Eq. (3). Since every descendant is either a correct copy of the parental genotype or a mutant, a conservation law holds: $\sum_{k=1}^{n} Q_{ki} = 1$. The *mutation matrix* $Q \doteq \{Q_{ki}\}$ is a stochastic matrix. In case of symmetric mutation frequencies, $Q_{ik} = Q_{ki}$, it is a bistochastic matrix.

Equation (5) can be solved explicitly, too.[20,33] As in Eq. (4) the solutions are superpositions of exponential functions, but now they are obtained as linear combinations of the eigenvectors $\vec{\ell}_j$ of a *value matrix* $W \doteq \{W_{ij} = A_j \cdot Q_{ij} - D_i \cdot \delta_{ij}\}$. The coefficients are time dependent: $u_j(t)$. The solution curves have the same general form as those shown in Eq. (4): the variables $x_k(t)$ have to be replaced by the coefficients $u_j(t)$ and and the eigenvectors of the value matrix, λ_j, appear in the exponential functions instead of the excess productions E_k.

The quasispecies Γ_0 can be defined now in precise mathematical terms. It is the combination of genotypes which is described by the dominant eigenvector of the value matrix, $\vec{\ell}_0 = (\ell_{10}, \ell_{20}, \ldots, \ell_{n0})$—this is the eigenvector which belongs to the largest eigenvalue: $\lambda_0 > \lambda_1 \geq \lambda_2 \geq \ldots \geq \lambda_n$. The quasispecies can be written as

$$\Gamma_0 = \ell_{10}\mathbf{I}_1 \oplus \ell_{20}\mathbf{I}_2 \oplus \ldots \oplus \ell_{n0}\mathbf{I}_n . \tag{6}$$

The components ℓ_{k0} measure the contributions of the individual genotypes $\mathbf{I}_k$ to the quasispecies. Their size is determined to a first approximation by the frequencies of mutation from the master sequence $\mathbf{I}_m$ to the genotype $\mathbf{I}_k$ and by the difference in selective values:

$$\ell_{k0} = \frac{Q_{km}}{W_{mm} - W_{kk}}, \quad k = 1, 2, \ldots, n \,; k \neq m \,; \; W_{kk} < W_{mm} . \tag{7}$$

Here we characterize the diagonal elements of the value matrix, $W_{kk} = A_k Q_{kk} - D_k$, as selective values. The master sequence is usually the genotype with the largest selective value—for exceptions see P. Schuster[30] and the next section.

Making use of the structure of matrix W it can be shown that all components of the eigenvectors $\vec{\ell}_0$ are positive. The sum of relative concentrations is constant, $\sum_{k=1}^{n} x_k(t) = 1$, and the same holds true for the sum of the coefficients of the eigenvectors, $\sum_{j=1}^{n} u_j(t)$. Accordingly the contributions of all eigenvectors except $u_0(t)$, that of the dominant eigenvector $\vec{\ell}_0$ vanish in the limit $t \to \infty$. After a sufficiently long time the mutant distribution converges towards the quasispecies.

The extension of the conventional selection model to the frequent mutation scenario has two important consequences:

- Goal of the selection process is not a fittest genotype but a clan of genotypes consisting of a master sequence—eventually of two or more master sequences—together with its frequent mutants. *Fitness* in the sense of Darwin's selection theory is a property of this clan, called the *quasispecies*. Fitness is *not* a property of the master sequence alone.

- Quasispecies in real—finite—populations become unstable at large mutation frequencies. Too many errors during replication accumulate and cause the loss of the master sequence and its mutants. Then, there exists no stable stationary mutant distribution.

The latter phenomenon is illustrated best in an abstract vector space of dimension ν—here ν is the length of the sequences. Those vectors which correspond to individual sequences are elements of the so-called *sequence space*.[17] Examples of low-dimensional sequence spaces of binary sequences are shown in Figure 2. Sequence spaces are identical to the configuration spaces of spin systems. In the case of sufficiently accurate replication the support of the population—this is the set of all points whose sequences are present in the population—is stationary. The mutant distribution is localized.[26] When the error frequency exceeds some critical value the population starts to walk randomly through sequence space. The surprising result of the theory is not the phenomenon as such, it is the sharpness of the transition from a localized *quasispecies* to a drifting mutant distribution which has much in common with phase transitions. It is well justified to speak of a critical error threshold at which the population becomes delocalized.

The relation between the error threshold of replication and phase transitions is much deeper than mere similarity. It was shown that statistical mechanics of

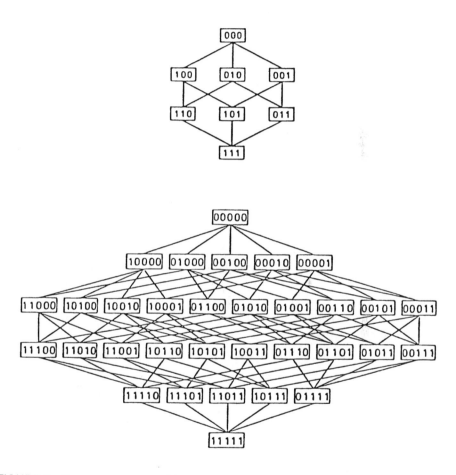

FIGURE 2 Sequence spaces of binary sequences of chain lengths $\nu = 3$ and $\nu = 5$. All pairs of sequences with Hamming distance $d = 1$ are connected by a straight line. The object drawn in that manner is a *hypercube* of dimension ν. The Hamming distance is the number of digits in which two sequences differ. Such a *vector-(-point-)-space* was used first by Richard Hamming[17] in information theory. The sequence space is identical with the configuration space of a spin system of the same dimension. The edges of the hypercube connect configurations which can be transformed into each other by a single spin flip.

equilibrium phase transitions in spin systems is equivalent to replication-mutation dynamics of binary sequences.[4,25] The localization-delocalization threshold of populations in sequence space corresponds to the order-disorder transition in spin systems. Examples are transitions from ferromagnetic or antiferromagnetic phases into the paramagnetic state.

Localization of quasispecies in sequence space is of fundamental importance for the evaluation of genotypes. Only localized populations conserve the information stored in sequences over sufficiently many generations. Only they can be evaluated by natural selection and are suitable for evolution. Delocalized, drifting populations cannot adapt evolutionarily to the environmental conditions.

A SIMPLE MODEL

Despite the enormous progress in the physical and chemical characterization of selection and evolutionary adaptation—which was achieved by means of the $Q\beta$ RNA replication essay—many questions cannot be addressed experimentally as yet. Among other problems there are the experimental determination of the distribution of selective values in sequence space and the analysis of mutant distributions—during adaptation and in the case of stationarity—which exceed by far the present-day capacities. Answers to these questions are, however, fundamental for an understanding of the evolutionary process. A simple model system was conceived which is suitable for numerical calculations and computer simulation.[10,32] In essence, the model is based on three assumptions:

- Binary (0,1) sequences are considered instead of real polynucleotides. As with RNA, non-neighboring elements of the polymer chain interact mainly through complementary base pairing, $0 \equiv 1$, and the stability of secondary structures is essentially determined by the number of base pairs.
- Variation of sequences is restricted to point mutations which exchange the two symbols at a given position in the polymer: $0 \to 1$ or $1 \to 0$. The chain length remains constant thereby.
- The accuracy of polymer synthesis—measured in terms of the frequency at which the correct symbol is incorporated into the growing polymer chain, $1 \geq q \geq 0$—is assumed to be independent of the position in the chain.

As a consequence of these three assumptions we can express the frequencies of all mutations $\mathbf{I}_i \to \mathbf{I}_k$—which are the previously mentioned elements of the mutation matrix Q—by the chain length ν, the Hamming distance of the two sequences, $d(\mathbf{I}_k,\mathbf{I}_i) = d_{ki}$, and a single parameter for the accuracy of replication q:

$$Q_{ki} = q^{\nu} \left(\frac{1-q}{q} \right)^{d_{ki}} . \tag{8}$$

The accuracy parameter of replication, q, is a measure of the precision with which individual digits are incorporated into the polymer chain: $q = 1$ implies error-free copying—every symbol in the copy is identical to that in the template. In the case $q = 0$ *always* the opposite symbol is incorporated—we are dealing with complementary copying which reminds one of the conventional photographic technique. Instead of positive and negative, *plus*- and *minus*-strands alternate here. Sequences occur in pairs, $(\mathbf{I}_k^{\oplus}, \mathbf{I}_k^{\ominus})$, in complementary replication. Another extreme case is observed at $q = 0.5$: both symbols, 0 and 1, are incorporated with equal probabilty—and independently of the symbol which stands at the corresponding position in the template. The error frequency is at maximum and there is no correlation between the sequences of template and copy. We may characterize this case appropriately as *random replication*.

Most simple distributions of kinetic parameters in sequence space are particularly well suited to illustrate some fundamental properties of quasispecies. We restrict selection to the synthesis of macromolecules and set all degradation rate constants equal: $D_1 = D_2 = \ldots = D_n = D$. The solutions of Eqs. (3) and (5) are invariant to additive constants and we choose $D = 0$ without losing generality. We consider here two examples of value landscapes for which the structures of quasispecies were studied as functions of the accuracy parameter q.[30,32]

THE SINGLE-PEAK LANDSCAPE

On the *single-peak landscape* a higher value of the replication parameter is assigned to the sequence $\mathbf{I}_0$, all other replication parameters are assumed to be equal: $A_0 = \alpha$ and $A_1 = A_2 = \ldots = A_{2^\nu - 1} = 1$. Consequently $\mathbf{I}_0$ is the master sequence ($m = 0$). The domain $1 \geq q \geq 0$ is partitioned into three ranges where direct, random and complementary replication dominate. The individual zones are separated by narrow transitions—even at chain lengths as short as $\nu = 10$. We may define two error thresholds, q_{min} for direct and q_{max} for complementary replication.

Direct replication dominates in the range $1 \geq q \geq q_{min}$. In the limit $q \to 1$ the quasispecies converges towards pure master sequence $\mathbf{I}_0$.

In the range $0 \leq q \leq q_{max}$ complementary replication dominates. At $q = 0$ a master pair consisting of $\mathbf{I}_0$ $(\mathbf{I}_0^{\oplus})$ and its complementary sequence $\mathbf{I}_0^{\ominus}$ is selected. For accuracy parameters $q > 0$ the master pair is accompanied by a distribution of its mutant pairs: $(\mathbf{I}_1^{\oplus}, \mathbf{I}_1^{\ominus})$, $(\mathbf{I}_2^{\oplus}, \mathbf{I}_2^{\ominus})$ etc. The goal of selection is a quasispecies of complementary pairs.

Chain elongation—increase in chain length ν—sharpens the two transitions remarkably. This can be understood as a strong hint that we are dealing with a cooperative process related to phase transitions. In Figure 3 we show the stationary mutant distribution for binary sequences of chain length $\nu = 10$ in the entire domain $1 \geq q \geq 0$. The stationary concentrations of the individual genotypes are essentially constant between the two critical values, i.e., in the range $q_{min} > q > q_{max}$. In addition, all seqeunces are present at the same frequency. This implies, however, that stationary solutions cannot be approached by realistic populations when the chain lengths exceed $\nu = 80$. No population can be larger than one-mol individuals, and $6 \cdot 10^{23}$ is roughly equal to 2^{80}. In case the population is too small, it cannot be localized in sequence space. Instead it drifts randomly and all genotypes have a finite lifetime only.

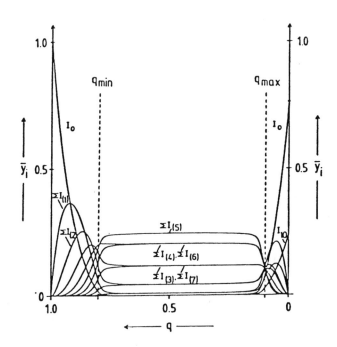

FIGURE 3   The stationary mutant distribution called *quasispecies* of binary sequences of chain length $\nu = 10$ on the *single-peak landscape* as a function of the replication accuracy q. A replication parameter $\alpha = 10$ was used. We plot the relative stationary concentrations of error classes as functions of q. Error classes are the master sequence I_0, all one-error mutants $I_{(1)}$, all two-error mutants $I_{(2)}$, ... , and finally the ν-error mutant $I_{(\nu)}$ of the master sequence. Two transitions occurring at critical values of the replication accuracy, q_{min} and q_{max}, are easily recognized. In the domain of random replication which lies between the two transitions, all individual genotypes are formed with essentially the same frequency—corresponding to a uniform distribution of sequences. Since we plot concentrations of error classes, the relative frequencies are proportional to the binomial coefficients in this range: $[I_{(k)}] = y_k = \binom{\nu}{k}/2^\nu$.

THE DOUBLE-PEAK LANDSCAPE

Two genotypes are characterized by high replication efficiency on *double-peak land-scapes*. In case these two sequences are close relatives—having a Hamming distance $d = 1, 2, \ldots$, up to some limit determined by population size, sequence length and structure of the value landscape—they are readily produced as mutants of each other and then a common quasispecies is formed at non-zero error rates, $q < 1$. It consists of the two highly efficient genotypes and a cloud of mutants surrounding both.

The case of more distant peaks appears more interesting. Then two distinct quasispecies—two distributions without common mutants—can exist. By an appropriate choice of the kinetic parameters both quasispecies can be stable. Each quasispecies has its own range of stability on the q-axis. These ranges don't overlap, instead they are separated by sharp transitions at critical replication accuracies q_{tr} which again resemble phase transitions.

A *double-peak landscape* which is well suited for computation and illustration of transitions between quasispecies puts the two highest selective values at maximal Hamming distance $d = \nu$—the two master sequences are complementary. Four different replication parameters are used for the two master sequences and their one-error mutants, in particular

1. $A_0 = \alpha_0$, for the master seqeunce $\mathbf{I}_0$,
2. $A_{(1)} = \alpha_1$, for all one-error mutants $\mathbf{I}_{(1)}$—we consider all one-error mutants together as an error class and express this by the subscript "(1),"
3. $A_{(\nu)} = \alpha_\nu$ for the second master sequence $\mathbf{I}_{(\nu)}$, and
4. $A_{(\nu-1)} = \alpha_{\nu-1}$ for all one-error mutants $\mathbf{I}_{(\nu-1)}$ of $\mathbf{I}_{(\nu)}$—these are the $(\nu-1)$-error mutants of $\mathbf{I}_0$.

The replication constant $A_k = 1$ is assigned to all other sequences.

An illustrative example of quasispecies rearrangement on a *double-peak land-scape* is shown in Figure 4. The two genotypes with high replication rates were assumed to be almost selectively neutral—the difference in replication rates is very small. The less-efficient genotype has more-efficient mutants than its competitor. Then the more-efficient master sequence dominates at low and the less-efficient one at high error rates.

The mechanism of the rearrangement of stationary mutant distributions at some critical replication accuracy, q_{tr}, is easily interpreted. The genotype with the somewhat higher replication parameter and the less-efficient neighbors in sequence space dominates at small error rates because the mutation backflow from mutants to master sequence is negligible. With increasing error rates—decreasing q-values—mutation backflow becomes more and more important, and at the critical replication accuracy, q_{tr}, the difference in replication rates of the two master sequences is just compensated by the difference in mutation backflow. Further increase in error rates makes the less-efficient master sequence superior to that with the higher replication rate constant. The conventional selection principle—as visualized by the rare mutation scenario—has to be modified in one aspect of primary importance:

■ Fitness is a collective property and it cannot be attributed to a single genotype. Fitness is determined by all sequences of the quasispecies and precisely to the extent in which they contribute to the stationary distribution.

Accordingly fitness is also an implicit function of the replication accuracy q.

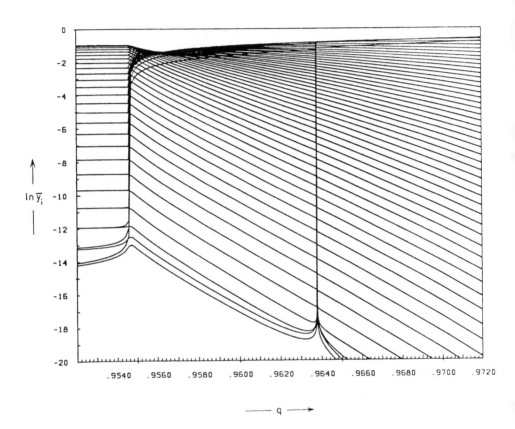

FIGURE 4 The stationary mutant distribution called *quasispecies* of binary sequences of chain length $\nu = 50$ on the *double-peak landscape* as a function of the replication accuracy q. Relative concentrations of the individual error-classes are plotted on a logarithmic scale in order to illustrate the sharpness of the two transitions. The two almost neutral master sequences are positioned at maximal Hamming distance, $d = 50$. The master sequence with higher replication efficiency, $\mathbf{I}_0$ with $\alpha_0 = 10$, is surrounded by less-efficient one-error mutants, $\alpha_1 = 1$, whereas the less-efficient master sequence, $\mathbf{I}_{(50)}$ with $\alpha_{50} = 9.9$, has better one-error mutants, $\alpha_{49} = 2$. Two sharp transitions are observed: a rearrangement of quasispecies at $q_{tr} = 0.9638$, where the two master sequences $\mathbf{I}_0$ and $\mathbf{I}_{(50)}$ exchange there roles, and the error threshold at $q_{min} = 0.9546$. $\mathbf{I}_0$ is the most frequent genotype in the range $1 \geq q > q_{tr}$ and $\mathbf{I}_{(50)}$ dominates in the neighboring range $q_{tr} > q > q_{min}$.

Studies on simple model landscapes are well suited for illustration of phenomena like error thresholds or rearrangements of quasispecies. They suffer, of course, from the fact that they have little in common with real value landscapes. In order to be able to proceed towards an understanding of selection and adaptation on the molecular level, it is inevitable to investigate more realistic models.

A MODEL OF A REALISTIC VALUE LANDSCAPE

Relations between genotypes and phenotypes are highly complex even in the most simple systems which are realized in test-tube RNA replication and evolution experiments. The phenotype is formed in a many-step process through decoding of the genotype. The outcome of this process is usually unequivocal in constant environment, but many genotypes may yield the same phenotype.

The genotype-phenotype relation in the model[11] considered here is reduced to the folding of a binary string into a secondary structure following a minimum free-energy criterion as described previously. In the actual computations we applied base pairing and base-pair stacking parameters of pure G,C-sequences.[12,35] The folding process is restricted to the formation of unknotted planar structures. This restriction is common in all presently used folding algorithms, and is also well justified on an empirical basis of the presently known structural data.

An important common feature of all correlations between genotypes and phenotypic properties is the *bizarre* structure of value landscapes. This means that closely related genotypes *may*—but need not—lead to phenotypes of very different properties. Distant genotypes, on the other hand, may yield almost identical phenotypic properties. This peculiar feature of genotype-phenotype relations is illustrated by means of secondary structures of binary sequences in Figure 5.

Rate constants of replication (A_k) and degradation (D_k) are particularly important properties in molecular evolution. The present knowledge on the dependence of these constants on RNA structures is not sufficiently detailed to allow reliable predictions. Nevertheless, it is possible to conceive models which reproduce qualitative dependence correctly. Such an ansatz was used in the computation of a value landscape[10] which describes the distribution of excess productions ($E_k = A_k - D_k$) in sequence space. In order to approach an optimal excess production, the replication rate should be as large as possible ($A_k \rightarrow$ max) and, at the same time, the degradation rate should be at the minimum ($D_k \rightarrow$ min). The two constants depend on secondary structures in such a way that both conditions cannot be fulfilled: the two trends contradict each other. An increase of A_k is commonly accompanied by an increase rather than a decrease in D_k. We thus observe conflicting structural dependences which remind of the frustration phenomenon in spin glasses and we expect a value landscape of high complexity.

0110001101001001101000001011010011000011101100101010001110010101000100

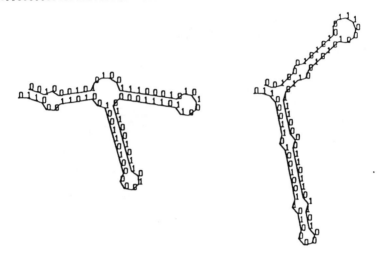

001011010001101010011011101100110000001010000000001010000101110000111011

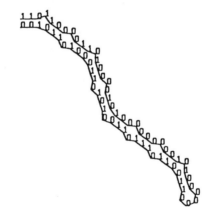

FIGURE 5 Examples of secondary structures of binary sequences. One main reason for the *ruggedness* of value landscapes is to be seen in the complexity of the folding process which presumably cannot be casted into any simpler algorithm. In the upper part of the figure we show two sequences of Hamming distance $d = 1$ which have very different secondary structures and, below that, two rather distant sequences with $d = 9$ which fold into identical secondary structures. Mutated positions are indicated by $*$.

In order to explore the details of the value landscape, two kinds of *computer experiments* were carried out:

- A population of 3000 binary sequences of chain length $\nu = 70$ was put on the value landscape in order to study the optimization by means of a genetic algorithm. The accuracy parameter was kept constant: $q = 0.999$. The stochastic dynamics of a chemical reaction network consisting of replication, mutation and degradation processes was simulated by means of an efficient computer algorithm.[15] Sequences produced in excess are removed by a dilution flux.
- The distribution of optimal and near-optimal regions of excess production (E_k) in sequence space was investigated by simulated annealing. In this search one- and two-error mutants were admitted as variations of the current strings.

The mean excess production $\overline{E}(t)$ increases steadily—apart from small random fluctuations—during an optimization run with the genetic algorithm and finally approaches a plateau value. The secondary structures obtained in three independent optimization runs which were performed with the genetic algorithm are shown in Figure 6. Despite identical initial conditions—3000 all-0-sequences of chain length $\nu = 70$—the three runs reach different regions in sequence space. These three regions of high excess production are far away from each other. The Hamming distances between these finally selected genotypes are roughly equal, and lie around $d = 30$. Although the kinetic properties of the three optimization products are very similar, they differ largely in sequence and secondary structure.

Depending on population size, replication accuracy and structure of the value landscape two different scenarios of optimization were found:

- The population resides for rather long time in some part of sequence space and moves quickly and in jumplike manner into another area. The optimal solution is approached by stepwise improvements.
- The population approaches the optimum gradually through steady improvement.

The first scenario is characteristic for small populations, low error rates and distant local maxima in sequence space whereas optimization dynamics of the second kind dominates in large populations, at high error rates and with lying close by local maxima. There is a steady transition from one scenario into the other: the gradual approach changes, for example, into a sequence of small jumps when the error rate is reduced. Larger jumps appear on further decrease of the mutation frequency and finally the population is *caught* in some local maximum.

Optimization runs with the genetic algorithm were also carried out at replication accuracies which lie below the critical accuracy for a population of 3000 molecules—the error threshold was found to depend on population size[28] and occurs at higher replication accuracies in smaller populations. No optimization on the

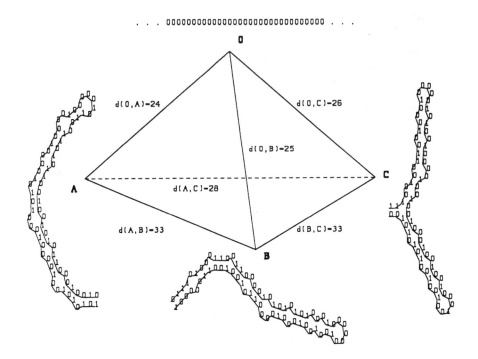

FIGURE 6 Secondary structures obtained in three independent optimization runs performed with the genetic algorithm. All three runs started with identical initial conditions, 3000 all-0-sequences of chain length $\nu = 70$, but used different seeds for the random-number generator. We show the most frequent genotypes at the end of the optimization processes. Interestingly, these three sequences and the all-0-sequence at the start have pairwise almost the same Hamming distances. The four sequences span an approximate tetrahedron in a three-dimensional subspace of the sequence space.

value landscape was observed. The populations drift randomly in sequence space. This general behavior was observed even when we started with *near-optimal* genotypes which had been optimized in previous runs. The initial master sequence is first surrounded by a growing cloud of mutants and then it is displaced by its own error copies. All other sequences in the population are lost likewise after a sufficiently long time. In other words, every genotype has only a finite lifetime. Therefore, the predictions of the molecular theory of evolution as introduced in the previous sections are valid also in populations as small as a few thousand molecules.

TABLE 1 The Hamming distances between the ten *best* sequences $(I_1, \ldots, I_{10})$ obtained by simulated annealing on the value landscape of the excess production (E_k) of binary sequences and their distance from the all-0-sequence (I_0)[1].

	I_1	I_2	I_3	I_4	Binary Sequences I_5	I_6	I_7	I_8	I_9	I_{10}
I_0	30	40	30	40	37	33	36	34	36	34
I_1		30	2	32	43	45	34	40	32	38
I_2			32	2	45	43	40	34	38	32
I_3				30	41	43	32	38	30	36
I_4					43	41	38	32	36	30
I_5						24	29	35	31	37
I_6							35	28	37	31
I_7								30	2	32
I_8									32	2
I_9										30

[1] The binary sequences are arranged in pairs of complementary and inverted sequeneces: I_2 is the inverted complementary sequence of I_1 ($I_2 = \neg I_1^\ominus$), I_4 is inverse complementary to I_3 ($I_4 = \neg I_3^\ominus$), etc.

In order to complement optimization by genetic algorithms, the value landscape was investigated by means of simulated annealing. The main goal of this study was to explore the distribution of high maxima of the excess production (E_k) and to compare the model value landscape of polynucleotide replication with a spin-glass landscape. The highest possible value of the excess production of binary sequences with chain length $\nu = 70$ can be computed from the criteria used in the evaluation of replication and degradation rate constants: it amounts to $E_{max} = 2245 \, [t^{-1}]$ in arbitrary reciprocal time units. Such a *best* structure, however, does not occur as a stable folding pattern. All sequences which, in principle, could yield the *best* structure fold into other secondary structures because of the minimun free-energy criterion—which of course need not meet a maximum excess-production criterion. The highest value of excess production obtained by simulated annealing was $E_{opt} = 2045 \, [t^{-1}]$. This optimal value is degenerate: it is found with ten distinct sequences. The Hamming distances between the ten sequences and the all-0-sequence are given in Table 1. The ten genotypes are related in pairs by symmetry: every sequence is transformed into one with identical structure and properties by complementation and inversion—inserting complementary symbols at all positions and swapping both ends (Figure 7). The remaining five sequences (I_1, I_3, I_5, I_7 and I_9) form two pairs,

$(\mathbf{I_1}, \mathbf{I_3})$ and $(\mathbf{I_7}, \mathbf{I_9})$, which are close relatives at Hamming distance $d = 2$ and one *solitary* sequence $(\mathbf{I_5})$. These potential centers for quasispecies are well separated in sequence space—all Hamming distances are larger than $d \geq 30$.

In total 879 sequences with excess production $E_k \geq 2011\,[t^{-1}]$ were identified. Their distribution in sequence space is shown in Figure 8 by means of a *minimum spanning tree*.[24] The clusters of peaks with high excess production have rich internal structures. High peaks are more probable in the neighborhood of other high peaks. Like in mountainous areas on the surface of the earth we find ridges as well as saddles and valleys separating zones of high excess production.

The distribution of near optimal configurations of the spin-glass Hamiltonian shows characteristic features of *ultrametricity*. Arbitrarily chosen triangles of three near-optimal configurations are either equilateral or isosceles with small bases. The distribution in sequence space of the high clusters of excess production, on the other hand, shows no detectable bias towards ultrametricity. This distribution of clusters does not deviate significantly from a random distribution. Two causes may be responsible for the difference between spin-glass Hamiltonians and value landscapes of polynucleotides. Either binary sequences with a chain length $\nu = 70$ are too short

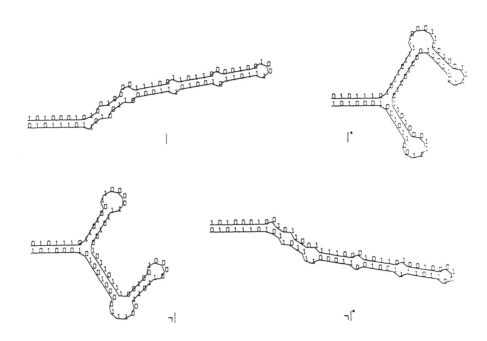

FIGURE 7 Symmetry of sequences with respect to secondary sturcture. Every sequence **I** at distance d from the all-0-sequence origin has a structure which is identical to that of the complementary and inverted sequence $(\neg\,\mathbf{I}^{\ominus})$ at distance $\nu - d$ from the origin.

to reveal higher-order structures in the value landscape or polynucleotide folding and evaluation of folded structures have some fundamental internal structure which is different from that intrinsic to the spin-glass Hamiltonian.

Finally, it is worth mentioning that optimization of excess production (E_k) by simulated annealing was not only faster—measured in CPU computer time units— but led also to significantly higher values than the genetic algorithm. This may be—at least in part—a result of the constant environmental conditions applied here. Constant enviroment favors algorithms with no or little memory. Storing of previously *best* solutions—as it is naturally done by populations—is advantageous only in variable environments.

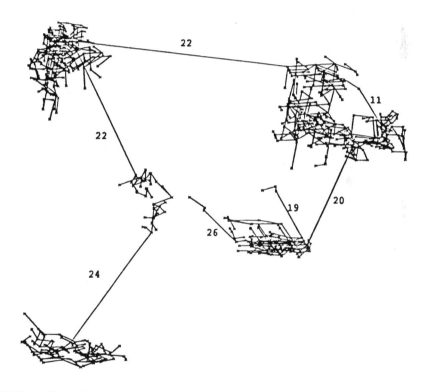

FIGURE 8   Clustering of near-optimal values of the excess production (E_k). The projection shows the *minimum spanning tree* of the frozen out optima of seven annealing runs operating under identical conditions (except random number seeds). The optimal regions reached by each run consist of an ensemble of selectively neutral mutants and qualify as clusters in the minimum spanning tree. The shortest Hamming distances between clusters are indicated.

CONCLUDING REMARKS

Optimization on a highly rugged value landscape derived from RNA secondary structures—modeled here by means of binary sequences—results in different scenarios depending on population size, error rate, and local details of the landscape. Extreme cases are the smooth and gradual approach towards a near-optimal solution and the approach in large steps, in which phases of stasis in optimization are interrupted by short periods of fast improvement. Intermediate cases were found as well. It is worth mentioning that the different scenarios occurred also in runs on different areas of the same value landscape with identical population sizes and error rates. Starting from low fitness values in *the planes* of the cost function, we observed the strong influence of the initial random walk on the final outcome: in the early phase of optimization the population proceeded always towards one of several areas of higher values—*the foothills*. The choice of the particular region that the population went to was done at random. This random choice, however, determined already some structural features of the final optimization products.

The rugged value landscape of our polynucleotide folding model has much in common with energy distributions in spin glasses. Apart from the apparent similarities there are, of course, substantial differences in the microscopic origins of complexity. In spin glasses, spin-spin coupling constants are randomly distributed—or they are at least much less regular than in solids—and this leads to *frustration*.[1] In biological evolution complexity can be traced down to the properties of polynucleotides. In the simplest cases it is the folding of strings into graphs which follows the rules of molecular biophysics—base pairing and stacking of base pairs such that they yield a minimal free energy—and which creates the complex structure of value landscapes. The notion of frustration applies here too: structural restrictions have to be fulfilled and therefore it is usually not possible to form the maximal number of base pairs.

Frustration appears again when folded structures are evaluated according to a maximum excess-production criterion ($E_k = A_k - D_k$). Here the origin of frustration is twofold. Firstly, according to model assumptions—certainly not met by reality— the genotypes which replicate fastest are also readily degraded. Genotypes which are particularly stable do not replicate very well. Optimization of excess production implies searching for a non-trivial compromise between these two trends.

The second source of frustration is typical for biological problems and was found—as a surprise—already in our simple model. The *best* structure having maximal excess production cannot be realized because of principal restrictions. All secondary structures have to meet a minimum free-energy criterion, and no genotype was found which can fulfil both criteria—maximum excess production and minimum free energy—at the same time. Boundary conditions restrict the wealth of possible solutions.

An important question concerns the significance of results derived from binary sequences for realistic RNA molecules representing (G,A,C,U)-sequences. Depending on the thermodynamic parameters used, binary sequences correspond either to pure (**G,C**)-sequences or pure (**A,U**)-sequences. In addition RNA molecules can be described by twice-as-long binary sequences. Each base is then encoded by two binary digits, for example by **G=11, A=10, U=01** and **C=00**. Complementarity relations hold—with the restriction that odd-numbered positions have to be opposite to odd-numbered positions in double-helical structures, and the same holds true for the even-numbered positions—but the evaluation of Hamming distances has to be modified and is not as straight as with binary sequences. The most important difference, however, concerns the richness in structures which is certainly larger with binary sequences. Therefore, point mutations in true RNA molecules will somewhat less likely lead to drastic changes in secondary structures than they do with binary sequences. Nevertheless we can expect that all qualitative findings reported here remain valid in four-base sequences.

The occurence of rugged value landscapes was found to be a fundamental property of biological macromolecules which can be traced down to the process of folding strings of digits into three-dimensional molecular structures. This fact is the ultimate cause, why transformation of genotypes into phenotypes is an exceedingly complex process—even in the most simple cases. Unfolding of genotypes cannot be reduced to simple algorithms therefore. Genotype-phenotype relation is one important—if not the most important—source of complexity in biology.

ACKNOWLEDGMENTS

The work reported here was supported financially by the *Fonds zur Förderung der wissenschaftlichen Forschung in Österreich* (Projects No.5286 and No.6864), the *Stiftung Volkswagenwerk* (B.R.D.), and the *Hochschuljubiläumsstiftung Wien*. Numerical computations were performed on the IBM 3090 mainframe of the *EDV-Zentrum der Universität Wien* as part of the IBM Supercomputing Program for Europe. In preparation of figures and reading of the manuscript by Dr. Walter Fontana and Mag. Wolfgang Schnabl is gratefully acknowledged.

REFERENCES

1. Anderson, P. W. *J. Less-Common Metals* **62** (1978):291–294.
2. Biebricher, C. K., and M. Eigen. "Kinetics of RNA Replication by $Q\beta$ Replicase." In *RNA Genetics, RNA-Directed Virus Replication*, edited by E. Domingo, J. J. Holland, and P. Ahlquist, vol. 1. Boca Raton, FL: CRC Press, 1988, 1–21.
3. Bounds, D. G. *Nature* **329** (1987):215–219.
4. Demetrius, L. *J. Chem. Phys.* **87** (1987):6939–6946.
5. Eigen, M. *Naturwissenschaften* **59** (1971):465–523.
6. Eigen, M., and P. Schuster. *Naturwissenschaften* **64** (1977):541–565.
7. Eigen, M., J. McCaskill, and P. Schuster. *J. Phys. Chem.* **92** (1988): 6881–6891.
8. Eigen, M., and C. K. Biebricher. "Sequence Space and Quasispecies Distribution." In *RNA Genetics: Variability of Virus Genomes*, edited by E. Domingo, J. J. Holland, and P. Ahlquist, vol. III. Boca Raton FL: CRC Press, 1988, 212–245..
9. Eigen, M., J. McCaskill, and P. Schuster. *Adv.Chem.Phys.* **75** (1989): 149–263.
10. Fontana, W., and P. Schuster. *Biophys. Chem.* **26** (1987):123–147.
11. Fontana, W., W. Schnabl, and P. Schuster. *Phys. Rev. A* **40** (1989):3301–3321.
12. Freier, S. M., R. Kierzek, J. A. Jaeger, N. Sugimoto, M. H. Caruthers, T. Neilson, and D. H. Turner. *Proc. Natl. Acad. Sci. USA* **83** (1986):9373–9377.
13. Garey, M. R., and D. S. Johnson. *Computers and Intractability: A Guide to the Theory of NP-Completeness.* San Francisco, CA: W. H. Freeman, 1979.
14. J. L. Gersting. *Mathematical Structures for Computer Science*, 2nd Ed. New York, NY: W. H. Freeman, 1987, 501–504.
15. Gillespie, D. T. *J. Comp. Phys.* **22** (1976):403–434.
16. Goldberg, D. E. *Genetic Algorithms in Search, Optimization and Machine Learning.* Reading, MA: Addison-Wesley, 1989.
17. Hamming, R. W. *Coding and Information Theory.*, 2nd Ed. Englewood Cliffs, NJ: Prentice Hall, 1986, 44–45.
18. Holland, J. H. *Adaptation in Natural and Artificial Systems.* Ann Arbor, MI: University of Michigan Press, 1975.
19. Hopfield, J. J., and D. W. Tank. *Biol. Cybern.* **52** (1985):141–152.
20. Jones, B. L., R. H. Enns, and S. S. Ragnekar. *Bull. Math. Biol.* **38** (1976):12–28.
21. Kauffman, S., and S. Levin. *J. Theor. Biol.* **128** (1987):11–45.
22. Kimura, M. *The Neutral Theory of Molecular Evolution.* Cambridge, U.K.: Cambridge University Press, 1983.
23. Kirkpatrick, S., and D. C. Gelatt, M. P. Vecchi. *Science* **220** (1983):671–680.
24. Kruskal, J. B. *Proc. Amer. Math. Soc.* **7** (1956):48–50.
25. Leuthäusser, I. *J. Stat. Phys.* **48** (1987):343–360.

26. McCaskill, J. S. *J. Chem. Phys.* **80** (1984):5194–5202.
27. Metropolis, N., and A. Rosenbluth, M. Rosenbluth, A. Teller, and E. Teller. *J. Chem. Phys.* **21** (1952):1087–1092.
28. Nowak, M., and P. Schuster. *J. Theor. Biol.* **137** (1989):375–395.
29. Rechenberg, I. *Evolutionsstrategie.* Stuttgart: Verlag, 1973.
30. Schuster, P., and J. Swetina. *Bull. Math. Biol.* **50** (1988):635–660.
31. Spiegelman, S. *Quart. Rev. Biol.* **4** (1971):213–253.
32. Swetina, J., and P. Schuster. *Biophys. Chem.* **16** (1982):329–353.
33. Thompson, C. J., and J. L. McBride. *Math. Biosc.* **21** (1974):127–142.
34. Zuker, M., and P. Stiegler. *Nucleic Acids Res.* **9** (1981):133–148.
35. Zuker, M., and D. Sankoff. *Bull. Math. Biol.* **46** (1984):591–621.

Somatic Evolution of Antibody Variable-Region Genes & Maturation of the Immune Response

Herman N. Eisen
Department of Biology and Center for Cancer Research, Massachusetts Institute of
Technology, Cambridge, MA 02139

Affinity Maturation: A Retrospective View

INTRODUCTION

This chapter is written in response to a request for an account of the work that
provided the first unambiguous evidence for systematic changes in the affinity of
antibodies synthesized in response to a single antigenic determinant. The impact
of the work derives in large measure, I think, from two considerations. First, it
provides a clear functional correlate for the somatic mutations that occur in the
rearranged immunoglobulin V genes of B lymphocytes. The concentration of these
mutations in hypervariable regions and the predominance of replacement over silent
ones, bear witness to the role of antibody affinity for antigen in the micro-revolution
of B-cell clones that attends responses to intensive immunization with protein and
protein-containing antigens. The second consideration is a matter of historical tim-
ing. When the work was published, it had a distinct bearing on the debate, still
not fully resolved at the time, between the two conflicting paradigms on the role of
antigen in antibody formation: the antigen-template theory and the clonal selection
hypothesis.

In the work at issue, published in 1964, Greg Siskind and I reported that a
2,4-dinitrophenyl (DNP)-protein antigen elicited in rabbits an evolving antibody
response to the DNP group. Antibodies isolated from early samples of antiserum

had low intrinsic affinity for the homologous hapten (ϵ-DNP-L-lysine), and those isolated from later samples had progressively increasing intrinsic affinity for the same hapten.[1] Later referred to as affinity maturation,[7] the progressive increase in affinity with time after immunization has come to be recognized as a general property of the antibody response to most protein antigens.

To appreciate the circumstances under which these observations were made it is useful to consider the conceptual framework in which the immune responses were viewed at the time. The major problem confronting immunologists in the 1950's was the structural and biosynthetic basis for the specificity of antibodies, a problem that in various forms has been recurrent and is still with us. Structurally, the question then focused on the need to understand how antibodies could be so similar to one another in overall structure and yet so different in specificity that there always seemed to be some antibody molecules that could react specifically with any of the virtually limitless number of different antigenic determinants (now termed epitopes) in the universe. The matching question was how this enormously diverse population of antibody molecules could be synthesized.

The antigen template theory seemed to provide a reasonably satisfactory answer to both questions, and it dominated the scene until around 1960. Proposed initially by Haurowitz and Breinl in the 1930's, this theory reached its full flowering in Pauling's compelling account of it in the early 1940's. Pauling's formulation made several predictions. One was that antibodies have two binding sites per molecule. The presence of multiple binding sites per antibody molecule had been proposed earlier by Heidelberger and Kendall as part of their lattice theory of the perception reaction. To simplify matters Pauling suggested minimal multivalency, i.e., bivalence. Another prediction was that the equilibrium association constants (affinity values) for all antibodies would, regardless of their specificity, fall within a limited range of values. This idea seemed reasonable since an antibody having an affinity that was below a critical threshold (say less than 10^4 liters per mol) would not bind strongly enough to the antigen to behave as an antibody, whereas with an affinity above some upper threshold (perhaps 10^6 liters per mol) it would bind too tightly to dissociate effectively from the template.

To measure antibody valence and affinity for a hapten, Fred Karush and I used equilibrium dialysis, a procedure that had been used earlier by Marrack, in England, to demonstrate that haptens would indeed bind specifically to antisera. As postdoctoral fellows in a Department of Biochemistry we qualified as rank amateurs and were unaware of Marrack's study (and, indeed, of almost everything else in the immunological literature). Karush was then using equilibrium dialysis to study alkyl sulfate binding to bovine serum albumin as a model system for analyzing protein-ligand interactions in general. I was interested in sulfonamide drugs as haptens and in raising antibodies against them to look for sulfonamide-protein conjugates in hypersensitivity lesions. In the course of working on these problems in the same laboratory, it became apparent that what Karush was doing with detergents and bovine serum albumin could be applied to haptens and antibodies. To pursue this possibility we chose the benzenearsenate-azo-system, because it had recently been extensively studied by Pauling, Pressman and Campbell. We immunized rabbits with benzenearsenate-azo-protein ("R-azo-protein"), isolated anti-benzenearsenate (anti-R) antibodies on R-azo red cell stroma (then in use as a solid support in a primitive form of affinity chromatography), and analyzed the interaction between the purified antibodies and a benzenearsenate hapten by equilibrium dialysis. The initial results were pretty spectacular. I remember well the contrasting appearance of deeply colored dialysis bags that contained the purified antibody (with bound and free hapten) and the pale dialysates (having free haptens only). By titrating the antibody with different initial hapten concentrations, it become clear that, at saturation, two moles of hapten were bound per mol of antibody, confirming the suggestion that antibodies were bivalent. It was also notable that these antibodies had the same specificity at each of their two sites, for as I recall it, there was much speculation at the time as to whether bivalent antibodies would have one site specific for the hapten and the second site specific for a protein epitope. Such antibodies of dual specificity were predicted by the antigen-template theory to be far more abundant than bivalent antibodies having the same specificity at each site, which is what our analysis revealed. But neither we[2] nor others could ever find antibodies of dual specificity, leading Felix Haurowitz, the originator and unyielding advocate of the antigen template theory, to reject the evidence for bivalent antibodies. He maintained for decades thereafter that antibodies are univalent.

The equilibrium dialysis binding curves, relating concentrations of bound to free hapten, were nonlinear. In line with a great deal of evidence accumulated over many years from many laboratories we attributed the nonlinearity to heterogeneity of the purified antibodies in respect to intrinsic affinity for the hapten. On the assumption that the distribution of affinities followed a normal distribution function, the "average" affinity was taken from the free hapten concentration when half the binding sites were occupied. In the samples that Karush and I studied initially the average intrinsic affinity was around 2×10^5 L/M. Karush subsequently studied other haptens, and we began to analyze antibodies to 2,4-dinitrophenyl (DNP) haptens. In all of Karush's studies and ours, and in those of Pressman and Nisonoff with other haptens, average binding constants of around 10^5 to 10^6 L/M were regularly found. The limited range fit well with what was expected from the antigen-template paradigm.

In our original work with the DNP hapten we followed procedures that were developed by Landsteiner et al., to couple aromatic amines via azo linkage to protein carriers, and we used the immunization schedules of the Heidelberger laboratory, in which large amounts of antigen (e.g., 25 mg per rabbit) adsorbed on precipitated alum, were injected. Subsequently, we found that 2,4-dinitrobenzenesulfonate introduced DNP groups exclusively into protein amino groups, and yielded highly substituted, stable, soluble proteins that were powerful immunogens when administered in small amounts in Freund's adjuvant. These changes had a pronounced affect on the amounts and properties of the antibodies produced. Earlier, antisera raised against DNP-azo-protein conjugates contained anti-DNP antibodies at a few hundred μg per ml at best; with the modified procedure the antibody concentrations were commonly over ten times higher.

The realization that these newly generated antibodies had extraordinarily high affinity grew out of studies carried out with Sidney Velick and Charles Parker.[9] Velick had shown that flavin and pyridine nucleotides could quench the tryptophan fluorescence of proteins that bound these ligands as coenzymes. The quenching was attributed to resonance energy transfer, and a condition for the transfer was spectral overlap between the protein's tryptophan fluorescence emission band and an absorption band of the bound ligand. We realized that these spectral conditions applied to antibody and DNP hapten, but it took some months of discussion before Velick and I could convince each other to carry out an experiment. The initial titrations with a purified anti-DNP antibody sample, prepared by the newer immunization procedures, and ϵ-DNP-L-lysine were striking. They were also surprising because they revealed that the antibodies have an average intrinsic association constant of about 5×10^8 L/M for ϵ-DNP-L-lysine, around 1000 times greater than the anti-DNP antibodies analyzed previously with the same hapten. To explain the disparity it was obvious that the effects of immunization conditions on antibody affinity had to be explored; two obviously important conditions were the time after immunization when antibody was harvested, and the amounts of antigen injected. The feasibility of carrying out a systematic exploration of these conditions was enormously enhanced by the logistical simplicity of the fluorescence-quenching method, for it required much less antibody and was carried out much more rapidly than equilibrium dialysis (e.g., only about 50 μg or less of antibody and about 10 minutes instead of a day to complete a titration).

Fortunately, Greg Siskind joined our lab as a postdoctoral fellow shortly afterwards and was able to carry out with enormous energy and effectiveness the extensive, systematic analysis required. It became clear that the anti-DNP antibodies isolated in the first week or ten days after DNP-antigen injection had an average intrinsic affinity of around 10^5 L/M, and those isolated subsequently increased progressively to where, after 1-2 months, the affinity was 10^8 L/M or greater (the upper limit measurable by fluorescence quenching). Two additional observations were made. (1) The rate of increase in affinity was greatly affected by the immunizing dose of antigen: with small doses, high-affinity antibodies appeared after a few weeks, but with high doses the appearance of these antibodies was greatly delayed. (2) All antibody samples were heterogeneous in respect to affinity, as expected, but it was also clear that the extent of heterogeneity increased with time, as the average affinity increased.

Before considering how all of these findings were interpreted it should be noted that changes in serum antibodies with time after immunization were not a new phenomenon. Aside from changes in isotype (from IgM, initially, to IgG later), it had been known for many years that antisera harvested at different times differed in their "avidity" for the antigen, the late antisera generally having higher avidity than early antisera.[6] Avidity, however, had a checkered history, because while it was (and is) commonly used by immunologists to refer to a tendency of antibodies and antigens to interact with each other, it has been defined only operationally, and in different ways by different investigators. Generally speaking, it refers, loosely I think, to the stability of antibody-antigen aggregates: those formed by highly avid sera dissociate more slowly than those formed by less avid antisera. However, various properties of the heterogeneous populations of antibodies in antisera can be responsible for these changes. For instance, one important property is the very diversity of these antibodies. A highly diverse antibody population is likely to recognize many epitopes of an antigen and to form a more stable aggregate with the antigen than a set of antibodies that recognizes only a few of the epitopes. In early serum samples, antibody populations are probably much less diverse than in late bleedings, and are expected, therefore, to form less stable complexes with antigens. On the other hand, increasing intrinsic affinity not only provided an explanation for the increase in avidity but suggested a more profound change in the evolution of the immune response, namely in the very binding sites of antibodies that are specific for the same haptenic group. The later finding was particularly critical for it came at a time when it could be related to the profound shift taking place in the way the immune system was viewed, i.e., from the antigen-template theory to the clonal selection hypothesis.

As I remember it, most biologically oriented immunologists seemed to accept clonal selection very soon after Burnet's seminal essay appeared in 1959, without waiting for confirmatory evidence. But more chemically minded immunologists gave up the antigen-template theory much more slowly and reluctantly. They (or should I say we?) probably would have abandoned the theory sooner if confronted with compelling evidence against it. But such evidence appeared only much later (as, for example, when it was shown by Haber[4] and by Tanford[10] that when antibodies are denatured and allowed to refold in the absence of antigen, they regain much of their original specificity for antigen). The evidence also appeared very slowly, and was more notable for its quantity—its sheer mass—rather than from the compelling quality of a single or even a few decisive papers. To be sure, John Humphrey and Gus Nossal reported around that time that they were unable to detect highly radioactive antigens in antibody-producing cells, but their results were hardly convincing to skeptics, because the limits of detection could not exclude the presence of some antigen that might have served as a template in these cells.

Taken at face value, the inverse relationship between antigen dose and the rate of appearance of high-affinity serum antibody was readily explained by the clonal selection hypothesis: the initial high levels of antigen would stimulate (i.e., select) clones making antibodies of diverse affinity whereas the low antigen levels present later in the response would selectively stimulate only high-affinity-producing cells. In contrast, the changes in antibody affinity, and the effect of antigen dose on the rate of change, were difficult for the antigen-template hypothesis to explain, unless the progressive increase in serum antibody affinity turned out to be an illusion resulting from a trivial secondary effect of the immunizing antigen. Thus, it could be (and was) argued that while the synthesized antibodies are diverse in respect to affinity, their average intrinsic affinity is invariant throughout the course of the entire response to an antigen. The average affinity of serum antibodies only appears to increase because of changes in the levels of the injected antigen. Shortly after antigen is injected, its tissue level would be high enough to bind and remove many of the newly synthesized antibodies, especially those with high affinity: the average affinity of the remaining ("free") antibody population in serum would thus be lower than the true average affinity. Later, as the antigen level declined, more of the high affinity antibodies would persist, resulting in an increase in average affinity of serum antibodies. Approached in this way, such a sequence would explain the persistence of antibodies of low average affinity for prolonged periods following immunization with unusually high doses of antigen, as Siskind and I had seen.

The alternative possibility, consistent with clonal selection, was that the antibodies synthesized at different times did indeed differ, with those synthesized later having binding sites with higher intrinsic affinity. Although a poll at the time would probably have shown that most immunologists were inclining towards clonal selection as an explanation, it seemed worthwhile to distinguish between the two possibilities.

The distinction was made by Lisa Steiner, then a postdoctoral fellow. She devised an elegant procedure for estimating the average relative affinity of the newly synthesized ^{35}S-methionine-labeled antibodies made by lymphocytes isolated from lymph nodes draining the site of antigen injection. Her results were decisive: the antibodies synthesized soon after a DNP-antigen had been injected had low average affinity for the haptenic group, whereas those synthesized weeks or months later had much higher affinity.[8] It was especially striking that after a long rest period following the initial immunization, a second injection of the same antigen promptly elicited a burst of synthesis of antibodies of extremely high affinity, at least as high as those in serum several months after the initial injection. These findings, especially those of the secondary ("booster") response, could not be accounted for by the antigen-template theory. But they were obviously consistent with selection by antigen among a population of B-cell clones having surface antibodies of diverse-affinities. The population seemed to undergo an antigen-driven change from the predominantly low-affinity antibody producers present initially to the predominantly high-affinity antibody producers that accumulated later in the primary response; the latter evidently persisted and were activated when the secondary response was subsequently triggered.

All of these interpretations rested on the heterogeneity of anti-hapten antibody binding sites, for which there was a vast amount of evidence extending back to the earliest work by Landsteiner et al. This heterogeneity, and the corresponding diversity of B cells, provided the basis for the antigen to selectively stimulate certain clones and not others. There were, however, those who questioned whether this antibody heterogeneity was inherent in the biology of the immune system or whether it was to some extent a reflection of non-uniformity in the hapten-protein conjugates commonly used as immunogens. Thus, in a typical conjugate such as DNP-bovine gamma globulin, the haptenic groups are attached to diverse lysine residues, forming in the same antigen molecule many different DNP-lysine-containing epitopes, each probably capable of eliciting a distinct and possibly homogeneous response. To pursue this possibility Haber et al. constructed some elegant antigens in which bradykinin was attached to polylysine in such a way that the bradykinin recurred as a uniform epitope.[5] They did, indeed, succeed in eliciting a relatively homogeneous antibody response against the haptenic group (bradykinin), but whether the responses were literally homogeneous was never established.

Using a different approach, we prepared a homogeneous immunogen by attaching a single DNP group to the particularly reactive lysine side chain at position 41 of bovine pancreatic ribonuclease (ϵ-41-DNP-ribonuclease). However, the anti-DNP antibodies elicited by this antigen turned out to be just as heterogeneous in respect to affinity for DNP-lysine as those elicited by conventional DNP protein conjugates. To me, this finding should have removed any lingering doubts that may have existed that the antibodies elicited against even a single epitope are heterogeneous. However, the view that heterogeneity of antibodies might arise from heterogeneity of the antigens continued to derive support from the remarkably homogeneous antibody responses to purified pneumococcal and streptococcal polysaccharides. As we subsequently learned, however, T cells do not recognize polysaccharides and the participation of T helper cells is required for an optimal B cell response, i.e., for affinity maturation.[3]

Of the three principal observations referred to above, the third and most confusing for some time had to do with the fact that heterogeneity with respect to affinity increased with time, as the average affinity increased. The prevailing speculation at the time was that the great diversity of B-cell clones resulted from somatic mutations in immunoglobulin genes. When, in the course of the development of B cells, did these mutations occur? The general view, I think, was that the mutations occurred early in B-cell development, before antigen was encountered, and indeed, that they were responsible for generating each individual's vast B-cell repertoire. Consequently we went through rather arcane arguments to try to explain why if high- as well as low-affinity clones were present in the preimmune animal, the initial response was so dominated by the low-affinity molecules. We suggested, among other things, that high levels of antigen would preferentially inactivate ("tolerize") the high-affinity clones: hence, only the low-affinity producers would be active early in the response, and if tolerance were reversible the high-affinity producers could recover and could be activated later, when antigen levels had declined. As it has turned out, however, somatic mutations occur predominantly after antigens stimulate immune responses. The activation of the mutation mechanism by immunization will probably turn out to provide a straightforward explanation for the progressive increase of affinity heterogeneity.

Some immunologists used to argue about whether some antigens were more physiological than others. These arguments have been rendered moot by the realization that the mutation-selection in B-cell clones is inherent in the optimal antibody response to all protein and protein-containing antigens. Whether the antigen is a natural constituent of a microbial pathogen or a protein carrying attached polysaccharides or small synthetic organic molecules, such as fluoresceine or DNP, is irrelevant. Each of these antigens acts as a selective agent, promoting the proliferation and maturation of those B cells that produce the antibodies that bind the antigen most effectively. In this sense, therefore, all antigens are equally significant biologically; the anti-DNP antibodies with which affinity maturation was discovered are just as physiological as the antibodies elicited by natural infection with an influenza virus. In looking forward to learning about the enzymes responsible for hypermutation and how they are regulated, one looks forward, in a sense, to a deepening understanding of how affinity maturation works.

REFERENCES

1. Eisen, H. N., and G. W. Siskind. "Variations in Affinities of Antibodies During the Immune Response." *Biochem.* **3** (1964):996.
2. Eisen, H. N., M. E. Carsten, and S. Belman. "Studies of Hypersensitivity to Low Molecular Weight Substances. III. The 2,4-Dinitrophenyl Group as a Determinant in the Precipitin Reaction." *J. Immunol.* **73** (1954):296.
3. Gershon, R. K., and W. E. Paul. "Effect of Thymus-Derived Lymphocytes on Amount and Affinity of Anti-Hapten Antibodies." *J. Immunol.* **106** (1971): 872–874.
4. Haber, E. "Recovery of Antigen Specificity after Denaturation and Complete Reduction of Disulfides in a Papin Fragment of Antibody." *Proc. Nat'l. Acad. of Sci. US* **52** (1964):1099.
5. Haber, E., F. F. Richards, J. Spragg, K. F. Austin, M. Valloten, and L. B. Page. "Modifications in the Heterogeneity of the Antibody Response." *Cold Spring Harbor Symp. Quant. Bio.* **32** (1967):299.
6. Jerne, N. K. "A Study of Avidity." *Acta Path. Microbiol. Scandinav. Suppl.* **87** (1951).
7. Siskind, G. W., and B. Benacerraf. "Cell Selection by Antigen in the Immune Response." *Adv. Immunol.* **10** (1969):1.
8. Steiner, L. A., and H. N. Eisen. "The Relative Affinity of Antibodies Synthesized in the Secondary Response." *J. Exp. Med.* **126** (1967):1143.
9. Velick, S. F., C. W. Parker, and H. N. Eisen. "Excitation Energy Transfer and the Quantative Study of the Antibody Hapten Reaction." *Proc. Nat'l. Acad. of Sci. US* **46** (1960):1470–1482.
10. Whitney, P. L., and C. Tanford. "Recovery of Specific Activity After Complete Unfolding and Reduction of an Antibody Fragment." *Proc. Nat'l. Acad. of Sci. US* **53** (1965):524.

C. Berek and M. Apel
Institute for Genetics, University of Cologne, Weyertal 121, D-5000 Cologne 41, F.R.G.

Maturation through Hypermutation and Selection

INTRODUCTION

Although the increase in affinity of the antibodies during an immune response was first described many years ago,[13] we are still far from understanding the cellular and molecular basis of the phenomenon. One fruitful approach developed in recent years has involved examining the change in antibody diversity during an ongoing immune response. These studies have demonstrated that the number of somatic mutations increases with time. This accumulation of somatic mutations correlates with an increase in antibody affinity for the antigen. Thus somatic hypermutation is an important factor in the maturation of the immune response. However hypermutation by itself is not sufficient. Assuming that the mutational mechanism operates more or less randomly over the entire variable region of the antibody molecules, many of the mutations will destroy rather than improve the affinity for the antigen. In order to see a maturation of the response, high-affinity variant B-cell clones have to be selectively expanded.

Molecular Evolution on Rugged Landscapes, SFI Studies in the Sciences of Complexity, vol. IX, Eds. A. Perelson and S. Kauffman, Addison-Wesley, 1991

THE RESPONSE TO 2-PHENYL-OXAZOLONE

One well-studied example of such a maturation process is the immune response to the hapten 2-phenyl-oxazolone (phOx). BALB/c mice were immunized with phOx coupled to the carrier chicken serum albumin (phOx CSA) and, at various time points thereafter, spleen cells were fused and hybridoma lines secreting phOx-specific antibodies isolated. The mRNA for the immunoglobulin molecules were directly sequenced and by this means the primary structure of the variable region of the H- and the L-chain determined.

Given the genetic mechanisms which ensure an enormous diversity in the variable regions of the antibody molecules, it is not surprising that there is a great variability in the primary structure of phOx-specific antibodies.[5] Virtually every phOx-binding antibody so far sequenced has a unique V(D)J region. Nevertheless, there are certain structural features which recur in the variable regions of most of the high-affinity antibodies. This implies that out of the broad spectrum available in the phOx-specific repertoire only a few B-cell clones are selected to further differentiate and to contribute towards the mature high-affinity immune response.

Primary immunization with the immunogen phOx CSA activates a subset of the available antigen-specific repertoire of B cells. Primary responses in general seem to start from naive B-cell clones whose receptors have not yet been diversified by the hypermutation mechanism. Many of these cells may differentiate into plasma cells, which leads to the early peak of antigen-specific IgM observed in the sera of immunized animals. These antibodies are rather diverse in their structure and often show only weak affinity for the antigen.[10] Nevertheless, these early IgM's may be quite important in that they provide a fast initial protection against the immunogen and by doing so give the immune system time to utilize its extraordinary capacity to perfect a "tailor-made" high-affinity response.

In contrast to the IgM antibodies the early primary IgG response to phOx is rather restricted.[7] The majority of the antibody molecules have one particular L-chain, $V_\kappa - Ox1$ joined to a $J_\kappa 5$ gene segment and an H-chain in which a $V_H - Ox1$ gene is combined with a characteristic D/J region. In these antibodies the D/J region is always 16 amino acids long. The D-segment consists of 3 amino acids, the first is invariably an Asp and the third invariably a Gly. It is on these $V_\kappa - Ox1/V_H - Ox1$ sequences that the first stages of the affinity maturation process will be centered. The low-affinity IgM-bearing B-cell clones appear not to contribute to the mature response. They may proliferate to some extent but they only rarely differentiate further. For instance, these low-affinity B-cell clones do not seem to switch to the IgG class.

Whereas 7 days after immunization only germline sequences were found, by day 14, sequences are detected carrying 2–3 somatic mutations per variable region.[6] In $V_H - Ox1$, mutations are more or less randomly distributed over the variable regions. In particular they are found in the complementary determining regions (CDR) as well as in the framework (FRW) residues. A strikingly different picture was obtained for $V_\kappa - Ox1$ L-chains where most of the sequences had mutations only in residues at the border of CDRI and FRWII. Chain recombination experiments have shown that mutations in CDRI (residue 34), an exchange of the germline encoded His to either Gln or Asn, increases the affinity 8- or 10-fold respectively.[3]

CORRELATION OF THE ONSET OF HYPERMUTATION AND THE FORMATION OF GERMINAL CENTERS

Over this early period of an immune response, morphological changes in the lymphatic system are taking place—changes which are as dramatic as the molecular events described above. About 4 days after immunization with a T-dependent antigen the germinal centers develop in the follicles of the lymphatic organs.[9] These substructures consist mainly of highly proliferating B cells. Germinal centers increase in size until day 10 and decline thereafter. The formation of germinal centers coincides with the time when the genes encoding antibody molecules are diversified by hypermutation and for this reason it was suggested that the microenvironment of the germinal center is necessary to activate the hypermutation mechanism.[8] In addition to B cells, large numbers of antigen/antibody complex presenting dendritic cells have been shown to be present in germinal centers. These cells may play an important part in the selection of high-affinity variants. In order to try to localize the maturation process we have analyzed antibody diversity in germinal-center B cells.

B-lineage cells within the germinal centers are characterized by their strong binding to a lectin, peanut aggultinin (PNA^{hi}). Thus the appearance of PNA^{hi} B cells after immunization correlates with the development of the germinal centers in the spleen of an immunized animal. By day 10 approximately 10% of the splenic B cells bind strongly to PNA. Using a fluorescence-activated cell sorter we have separated PNA^{hi} B cells from those which stain only weakly (PNA^{lo}). Both fractions were fused and hybridoma lines isolated.[1] Ten days after immunization 90% of the antigen specific hybridoma lines were derived from the PNA^{hi} B-cell population. These results suggested that by day 10 practically all of the activated B cells are in the germinal centers. Only a few antigen-specific hybridoma lines were obtained either from the PNA^{hi} or PNA^{lo} populations 13 and 14 days after immunization. This lower yield of antigen specific lines indicates that the numbers of activated B cells has declined two weeks after primary injection of antigen. In accordance with this finding the number of germinal-center B cells have started to decrease in the spleen of immunized animals.

ANTIBODY DIVERSITY IN DAY-10 GERMINAL-CENTER B CELLS

Antibody diversity was analyzed in hybridoma lines derived from day-10 germinal-center B cells by directly sequencing the mRNA for the H-chain and the L-chain. The expressed repertoire of the germinal-center cells reflects what has been observed in IgG antibodies derived from fusions using total spleen cells of immunized animals. For instance, in a fusion using PNA^{hi} cells of a BALB/c mice, 64% of the hybridoma lines had the $V_H/V_\kappa - Ox1$ combination. However, in contrast to results obtained from total spleen cells analyzed 7 days after immunization,[7] this dominance was found for IgM as well as for IgG antibodies. The data show that only a selected population of B cells starts to proliferate in the germinal centers. Low-affinity B-cell clones are somehow excluded.

Most of the lines (12 out of 16) carried somatic mutations. However the pattern of mutations in these sequences is different from the pattern observed at later stages of the immune response. Ten days after immunization there is no evidence for a preferential expansion of high-affinity variants; the characteristic substitutions found at the border of CDRI to FRWII of $V_\kappa - Ox1$, which are known to increase the antibody affinity, were not seen in any of the $V_\kappa - Ox1$ L-chains (Figure 1). The pattern of somatic mutations in the day-10 sequences reflects the situation prior to positive selection by antigen.

One way to determine whether positive selection has occurred during the expansion of B-cell clones is to calculate the ratio of replacement to silent mutation (R/S value) for the CDR and the FRW regions of the antibody molecules.[14] Table 1 shows the R/S values obtained for the $V_\kappa - Ox1/V_H - Ox1$ sequences at different time points of the immune response to the antigen phOx. Ten days after immunization, at the onset of hypermutation, the majority of the mutations are silent, resulting in low R/S values. Not only the FRW but also the CDR regions of the $V_\kappa - Ox1$ sequences gave R/S values which are lower than one would expect by a random distribution of mutations (R/S 2,9). From this result one may conclude that not only are replacement mutations selected against in the FRW but also in the CDR. This notion is supported by the recovery of a PNA^{hi}-derived B-cell line expressing the typical $V_\kappa - Ox1/V_H - Ox1$ combination but which failed to bind antigen. This line carries a replacement mutation in CDRIII in a residue believed to be important for the binding of phOx (line H23/7, Figure 1).[6]

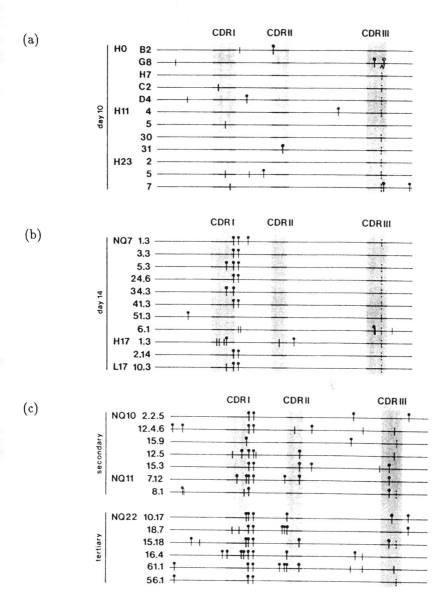

FIGURE 1 Somatic diversification of $V_\kappa - Ox1$ sequences. (a) Onset of somatic hypermutation; (b) selected mutations; and (c) accumulation of mutations. A diagrammatic comparison of $V_\kappa - Ox1$ mRNA sequences, where lines H0/C2 and H0/D4 express V_κ-genes related to $V_\kappa - Ox1$,[1] is shown: sequences from germinal-center B cells (a), from the day-14 immune response (b), from the secondary and the tertiary response (c) are given. Data are taken from the literature.[1,2,4,6] A vertical bar shows the relative position of codons containing a nucleotide difference, a black circle denotes a replacement mutation, and |♀ indicates junctional diversity and ∧ an insertion of an additional Pro at the border of V_κ to J_κ. Complementarity-determining regions (CDRI, II, and III) and J- region have been marked.

ANTIBODY DIVERSITY 14 DAYS AFTER IMMUNIZATION WITH ANTIGEN

Fourteen days after immunization two different types of sequence were obtained: those with a few selected mutations and those with many—predominantly silent—mutations (Figure 2). Highly mutated lines were recovered twice, once from a fusion of PNA^{hi} cells (Figure 2, line H17/1.3) and once from a fusion of bulk spleen cells (Figure 2, line NQ7/6.1). Both fall into the same category as day 10 sequences in having a low ratio of replacement to silent mutations (Table 1, day 14$^-$). In the $V_\kappa - Ox1$ sequence of antibody H17/1.3 one of the silent mutations has occurred at the same position as in two lines derived from day-10 germinal-center B cells. This means that in nine mutated $V_\kappa - Ox1$ sequences which have been obtained from PNA^{hi} cells, a silent mutation is found three times at the same position. Similarly out of six mutated $V_H - Ox1$ sequences, the same silent mutation was found twice. These results suggest that the silent mutations are not randomly distributed over the $V_\kappa - Ox1/V_H - Ox1$ variable regions. Mutations seem to accumulate at positions which might be hot spots for somatic mutations.

Most (8 out of 11) of the day-14 sequences derived from fusions[1] of PNA^{hi} and bulk spleen cells[6] showed only few mutations (1–4) per variable region. The majority of these mutations were replacement mutations. Particularly in the $V_\kappa - Ox1$ L-chains, high R/S values were found (Table 1, day 14), pointing to a strong antigenic selection for B cells expressing mutations which increase the affinity for antigen. In contrast to results obtained with a variety of different H- and L-chain, variable regions[12] in the $V_\kappa - Ox1$ L-chain R/S values for CDR and FRW are high (Table 1, day 14); in most sequences mutations were found in residue 34 of CDRI and residue 36 of FRWII. Structural analysis of antibody molecules suggests that in certain V regions the CDR may be extended and thus come to include residues which are normally considered as part of the FRW.[11]

TABLE 1 Ratio of replacement (R) to silent (S) mutations

| | $V_\kappa - Ox1$ | | | | | | $V_H - Ox1$ | | | | | |
| | CDR | | | FRW | | | CDR | | | FRW | | |
	R	S	R/S	R	S	R/S	R	S	R/S	R	S	R/S
day 10	3	3	1.0	3	4	0.7	2	-	≥ 2	2	3	0.7
day 14^{sel}	10	1	10	9	-	≥ 9	4	2	2.0	6	6	1.0
day 14$^-$	2	5	0.4	1	3	0.3	-	1	≤ 1	3	3	1.0
sec	14	5	2.8	13	4	3.3	6	-	≥ 6	1	2	0.5
tert	17	3	5.7	15	3	5.0	8	1	8.0	8	5	1.6

ACCUMULATION OF SOMATIC MUTATIONS IN THE SECONDARY AND TERTIARY RESPONSE

Nearly all $V_\kappa - Ox1$ sequences of the secondary (NQ10 and NQ11)[2] and the tertiary response (NQ22)[4] carry the characteristic mutations at positions 34 and 36 of the $V_\kappa - Ox1$ L-chain. In addition many more mutations have accumulated (Figure 3). However it may be that it is difficult to further improve the affinity of the antibody $V_\kappa - Ox1$ L-chain by the accumulation of further mutations in $V_\kappa - Ox1$. The R/S values of secondary- and tertiary-response antibodies are lower than at day 14 suggesting that most of the additional mutations are neutral and do not improve the affinity for the antigen. In contrast in $V_H - Ox1$ variable regions the R/S values in the CDRs increase from day 14, through the secondary and tertiary response. The overall increase in affinity of tertiary-response antibodies may be caused by improvements in $V_H - Ox1$.

MATURATION OF THE IMMUNE RESPONSE

Analysis of day-10 germinal-center-derived hybridoma lines suggests that the hypermutation mechanism is active when B cells proliferate in the follicles of the lymphatic tissue.[1] Nucleotide exchanges are introduced into the V-regions of the antibody molecules with a high rate. They seem to accumulate at positions which might be hot spots for somatic mutation. The preponderence of silent mutations which we see in antigen-specific hybridoma lines indicates that many of the replacement mutations both in FRW and in the CDR are deleterious for the antibody molecules.

Only four days later, a rather different pattern of somatic mutations is seen, particularly in the $V_\kappa - Ox1$ sequences. By and large day-14 $V_\kappa - Ox1$ L-chains have no silent mutations but all carry the typical mutations at residues 34 and 36 which are known to increase the affinity. One explanation for this would be that during the primary response the hypermutation mechanism is activated for only a short period and therefore only a few somatic mutations are introduced per variable region. Receptors which by chance have picked up mutations with increasing affinity for the antigen are than selectively expanded without accumulating further somatic mutations. One might envisage a model in which high-affinity binding of antigen to the receptor gives a signal to the cells to inactivate the hypermutation process.

Expansion of B-cell clones without further accumulation of mutations would, by itself, give these cells a proliferative advantage. A preferential expansion of high-affinity variants would be ensured since such cell lines would not pick up secondary—possibly deleterious—mutations.

As one progresses from the primary to the secondary and the tertiary responses, the number of somatic mutations increases, suggesting that the activation of memory cells by antigen reactivates the hypermutation mechanisms. At the same time the overall affinity for the antigen increases. This means that the selective mechanism which have led to the preferrential expansion of high-affinity variant clones are also active at later stages of the immune response. Therefore, if the formation of germinal centers is necessary for the activation of the hypermutation mechanism and the selection of high-affinity variants, one would predict that memory cells re-enter the follicles and form germinal centers upon activation with antigen.

ACKNOWLEDGMENT

We would like to thank U. Weber for technical assistance and U. Ringeisen for graphical work.

REFERENCES

1. Apel, M., and C. Berek. "Somatic Mutations in Antibodies Expressed by Germinal center B-Cells Early After Immunization." *Int. Immunol*, 1990, in press.
2. Berek, C., G. M. Griffiths, and C. Milstein. "Molecular Events During Maturation of the Immune Response to Oxazolone." *Nature* **316** (1985):412–418.
3. Berek, C., and C. Milstein. "Mutation Drift and Repertoire Shift in the Maturation of the Immune Response." *Immunol. Rev.* **96** (1987):23–41.
4. Berek, C., J. M. Jarvis, and C. Milstein. "Activation of Virgin and Memory B-Cell Clones in Hyperimmune Animaly." *Eur. J. Immunol.* **17** (1987):1121–1129.
5. Berek, C., and C. Milstein. "The Dynamic Nature of the Antibody Repertoire." *Immunol. Rev.* **105** (1988):1–26.
6. Griffiths, G. M., C. Berek, M. Kaartinen, and C. Milstein. "Somatic Mutation and the Maturation of the Immune Response." *Nature* **312** (1984):271–275.
7. Kaartinen, M., G. M. Griffiths, A. F. Markham, and C. Milstein. "mRNA Sequences Define an Unusually Restricted IgG Response to 2-Phenyl-Oxazolone and Its Early Diversification." *Nature* **304** (1983):320–324.
8. MacLennan, I. D. M., and D. Gray. "Antigen Driven Selection of Virgin and Memory B-Cells." *Immunol. Rev.* **91** (1986):61–85.
9. Nieuwenhuis, P., and D. Opstelten. "Functional Anatomy of Germinal centers." *American J. Anatomy* **170** (1984):421–435.
10. Pelkonen, J., M. Kaartinen, and O. Mäkelä. "Quantitative Representation of two Germline V-Genes in the Early Response to 2 Phenyloxazolone." *Eur. J. Immunol.* **16** (1986):106.

11. Poljak. Personal communication, August 1989.
12. Shlomchik, M. J., S. Litwin, and M. Weigert. "The Influence of Somatic Mutation on Clonal Expansion." *Prog. Immunol.* **7** (1989):415–423.
13. Siskind, G. D., and B. Beanceraff. "Cell Selection by Antigen in the Immune Response." *Adv. Immunol.* **10** (1969):1–50.
14. Weigert, M. "The Influence of Somatic Mutation on the Immune Response." *Immunol.* **6** (1986):139–144.

Catherine A. Macken† and Alan S. Perelson‡
†Department of Mathematics, Stanford University, Stanford, CA 94305 and ‡Theoretical
Division, Los Alamos National Laboratory, Los Alamos, NM 87545

Affinity Maturation on Rugged Landscapes

1. INTRODUCTION

One of the fundamental modes of response of the body to invading antigen is the
secretion of antibody by B-lymphocytes. In order to explain the ability of the immune system to make antibody which could bind essentially any antigen, Burnet[7]
proposed the clonal selection hypothesis. According to this hypothesis a diverse
population of B-lymphocytes exists, with each lymphocyte committed to the expression of a single antibody structure. Prior to antigenic challenge, B-lymphocytes
express antibody as cell surface receptors. When antigen binds and cross-links these
surface receptors, the cell can be stimulated to proliferate and differentiate into an
antibody secreting state, thus leading to a quantitative increase in antigen-specific
antibodies.

Recently it has become apparent that there is additional complex fine-tuning involved in antibody-mediated immunity. When antibody secreting cells are examined
at various times during primary and secondary immune responses, it is apparent
that the genetically encoded antibody structures are progressively and extensively

altered as a result of somatic mutation in the genes coding for the variable regions of the antibody heavy and light chains. In conjunction with these accumulating mutations, an increase in the average affinity of antibody for the stimulating antigen is found. This phenomenon is called *affinity maturation*, and in this book is the subject of many of the chapters including the present one.

Many intriguing questions exist about the detailed mechanisms of affinity maturation. We have developed a mathematical model of those aspects of the maturation process attributable to the random nature of somatic mutation and selection for fitter variants. In the absence of a model, processes that exhibit randomness are difficult to study. One is never quite sure if a particular observation is characteristic of the process or if it is a rare event unlikely to be repeated. Thus, an empirical approach of inferring mechanisms from a finite, possibly small, set of observations is prone to error. Mathematical models operate in the opposite direction to empirical analyses; they postulate mechanisms and then use mathematics to determine the consequences of the postulates. The mathematical model we develop here uses the notion of a *rugged landscape*. Our modeling approach illustrates the way in which imagery evoked by the term "rugged landscape" can be translated into quantitative predictions about the affinity maturation process. As we will show, our model makes a set of predictions, many of which are matched closely by experimental evidence, and others of which have not yet been tested.

2. MODEL

During the course of an immune response the average *affinity* or equilibrium binding constant of antibodies for the immunizing antigen increases with time. This increase of affinity has been studied experimentally in more detail by making hybridomas from antibody-secreting cells, measuring the affinity of the resulting monoclonal antibody, and sequencing the immunoglobulin mRNA (cf., Manser, this volume; Berek and Apel, this volume). B-cell clones that grow large enough to be detected[1] are noted to have mutated from the germline V-region sequence by a so-called *somatic hypermutation* process, in which point mutations are introduced into those chromosomal regions containing antibody V-region genes at a rate estimated to be 10^{-3} per base pair per generation (cf., McKean et al.[26]). In conjunction with the accumulation of mutations, antibodies tend to have an increased affinity for the immunizing antigen. Increases in affinity of order ten- to fifty-fold are common, while larger increases are much rarer. After an initial improvement in affinity of an order of magnitude or so, further point mutations tend not to lead to additional substantial improvements. It is typical to observe 6-to-8 point mutations leading to amino acid replacements, although the number varies greatly from antibody

---

[1]The frequency of hybridoma formation from activated B cells is low, perhaps only 1 in 1,000 cells (Bruggeman et al.[6])

to antibody. Higher affinity antibodies are observed in secondary responses, but these antibodies are usually derived from germline V genes that differ from the germline genes utilized in the primary response. The effects on affinity of particular mutations can be studied by site-directed mutagenesis. It has been suggested that each selected mutation leads to an increase in affinity; thus affinity may be viewed as increasing in a step-wise fashion.[13,20,32]

The theoretical framework, described below, involves viewing the molecular evolutionary process as analogous to walking on a rugged landscape. We are encouraged with the potential appropriateness of our preliminary theory since we predict 8 ± 5 mutations (mean $\pm$ 2 s.d.) in V regions of antibodies that have improved in affinity from their germline antibodies. This prediction roughly corresponds to the range seen in experiments. Although our theory is not correct in detail, we believe it captures some of the important features of the molecular evolutionary process. For example, it provides an explanation for the slowing of the rate of affinity improvement with successive mutations, and for the fact that rare V-region genes can ultimately provide high affinity antibodies.

The method of analysis that we use is general and not restricted to antibody evolution, although it is best suited to the study of evolution by point mutation as opposed to recombination. In our theory, conceived in collaboration with Stuart Kauffman of the University of Pennsylvania and analyzed in collaboration with Patrick Hagan of Los Alamos National Laboratory, we represent a protein or nucleic acid and all of its mutant forms of the same length as points in an abstract *sequence space*. To each sequence we assign a fitness and then allow evolution to occur by steps of increasing fitness. This method, first suggested by John Maynard-Smith,[25] was later used and refined by Jacques Nino,[27] Manfred Eigen,[10] John Gillespie,[14] Zvia Agur and Michael Kerszberg,[2] Stuart Kauffman,[16,17,19,34] and Peter Schuster,[12,30,31] and their coworkers. The method thus has three components: a sequence space, a function which assigns a fitness to each sequence, and a "move rule" which determines how one sequence is converted to another by evolution. We now discuss each component in turn.

In the model, we describe an antibody variable (V) region by a sequence of N symbols, each symbol being chosen from an alphabet of a letters. If antibody V regions are viewed at the protein level, then $a = 20$ and N is approximately 230, the number of amino acids in both the heavy and light chain V regions. If we view antibody V regions at the level of DNA, then N is approximately 700 and $a = 4$. Our theory is independent of the mode of describing antibodies. The set of all a^N possible configurations of sequences of length N constitutes *sequence space*. Two sequences that differ in one position only are called *one-mutant neighbors*. Two-mutant and j-mutant neighbors can be defined analogously. Since a single point mutation changes one letter in a sequence, evolution by point mutation can be viewed as a connected walk among one-mutant neighbors in sequence space. The number of one-mutant neighbors of a sequence is defined to be D. In protein design experiments, $D = 19N$, since each possible amino acid substitution creates a unique one-mutant neighbor. However, for proteins, such as antibodies, evolving by point mutations in DNA, we replace the actual number of one-mutant neighbors, $3N$, where N is the number of base pairs, by an effective number of one-mutant neighbors at the DNA level which is less than the maximum possible. Because of restrictions in the genetic code, not all point mutations lead to a change in protein composition. On average, only 75% of DNA point mutations are expressed as amino acid changes. Thus for an antibody V region composed of 700 base pairs, the effective number of one-mutant neighbors $D \simeq 0.75 \times 3 \times 700 = 1575$. To indicate that this value of D is only approximate we shall assume $D = 1500$ for antibodies. The quantitative predictions of our theory will turn out to depend only on the logarithm of D, so it is somewhat immaterial if D is 1500 or 1600. By using this reduced value of D we restrict our attention to replacement mutations that can affect the fitness of the antibody.

To each sequence in sequence space we need to assign a fitness. A natural measure of fitness, which we adopt here, is the antibody affinity for the immunizing antigen. Although the affinity of each antibody sequence can be determined experimentally, this is clearly not feasible given the large number of possible sequences. Thus one would like a means of predicting affinity from sequence. To do this, one would probably need to predict the three-dimensional structure of both the antibody combining site and the antigen, and then solve what is known as the "docking" problem to determine the interaction of the two molecules. At the moment this is also not feasible. In the absence of a structurally based calculation of fitness, we adopt a simpler, although perhaps less precise, approach. We note that examination of naturally occurring mutations and mutations generated by site-directed mutagenesis shows that changing some amino acids in an antibody has little effect on affinity, whereas changes in other amino acids can lead to an order-of-magnitude increase in affinity, a decrease in affinity, or complete loss of affinity.[8,28,29] One may speculate that certain regions of the antibody, such as the framework regions, are extremely sensitive to mutational changes, and are therefore highly conserved, while the complementarity-determining regions can be mutated more freely. In the current model, we do not distinguish between regions of sensitivity to mutation. We simply state that affinity cannot be predicted from sequence with certainty, and hence assign each antibody a fitness randomly selected from a specified probability distribution. The distribution may be normal, lognormal, uniform, etc. If one chooses a distribution that is tightly peaked around its mean, then one is assuming that a single point mutation will most likely not change affinity very much. Conversely, using a uniform distribution implies that a single mutation has equal probability of generating any allowed affinity.

We do not believe that fitnesses are random functions of sequence. However, this choice of relationship provides information about one extreme possibility. If fitnesses are random, neighboring sequences can have very different fitnesses, and the landscape will be *rugged*. Even if the fitness function is not random the landscape can be rugged. Thus a study of random fitness functions provides qualitative information about the class of rugged landscapes. A similar philosophy has proven valuable in physics, where the study of systems with random energy functions has lead to insights into the properties of spin glasses (see chapters by Palmer and Stein).

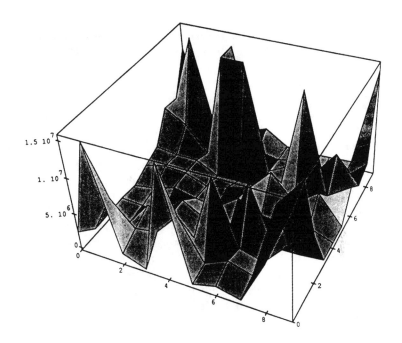

FIGURE 1   A schematic illustration of a rugged fitness landscape based on 100 sequences. The sequences are arranged in a 10 x 10 grid and the fitnesses are represented as a height above each grid point. The fitness values were chosen randomly from a lognormal distribution, Eq. (2), with $\mu = 6$ and $\sigma = 2/3$.

Once a fitness is randomly assigned to each sequence, a landscape is generated, with the fitness being the height of the landscape. Figure 1 is an attempt to illustrate a rugged landscape for $N = 2$ and $a = 10$, with fitnesses chosen from a lognormal distribution. In the figure, 100 sequences are represented as points in the plane, (x_i, y_i), $i = 1, \ldots, 100$, where x_i and y_i are the amino acids (letters) in positions 1 and 2 of the ith sequence. To each sequence, a fitness is assigned and plotted as the height of the surface above the plane, i.e., z_i is the fitness of sequence i. This figure is not a precise representation of the rugged landscapes that we study in this paper because we cannot represent in three dimensions the important property that each sequence has $D = N(a - 1) = 18$ one-mutant neighbors. For example, the one-mutant neighbors of the sequence at point (4,0) in the figure are all the sequences with the same x or y coordinate, i.e., the sequences $(4, j)$, $j \neq 0$ and $(k, 0)$, $k \neq 4$. The problem with Figure 1 is that sequence space is represented as a square lattic and thus each sequence has at most four neighbors. A much higher dimensional space, which we obviously cannot draw, would be needed to represent a space in which each point has 18 equidistant neighbors. A further property of a rugged landscape, which is pictured in a particularly misleading fashion, is the number of local optima. One needs to realize that a sequence which is a local optimum must have a fitness which is higher than that of all of its 18 one-mutant neighbors, not just the four represented here.

Maynard-Smith[25] suggested that protein evolution generally occurs by means of single mutations leading to a higher fitness. Proteins differing by two mutations with an intermediate configuration having a lower fitness than both of the end configurations cannot occur frequently since the intermediate protein will be expressed at such a low rate in the population that the next mutation is unlikely to take place. By adopting Maynard-Smith's suggestion that selection operates to retain only mutations of higher fitness, we obtain a rule by which we move in sequence space, i.e., we move only in a direction of increasing fitness.

Before formally discussing the predictions of our model, we consider the relationship between the model and the biological processes we wish to study. In order to do this, let us consider in greater detail the way in which we model evolution as a walk on a landscape. The germline sequence defines a starting point on the landscape. At any stage on a walk, single-mutant variants of the current antibody are tested in random order until the first neighbor having a higher fitness is attained. The walk then moves to this new point in sequence space (i.e., new antibody sequence), and the testing process starts anew. Thus, each move on a walk changes the current antibody sequence to a neighboring one. As there are D one-mutant neighbors of each point in sequence space, at most D different single mutant variants can be tested. If no fitter variant is found among these different one-mutant neighbors, the process stops, as the walk has reached a local optimum.

There are two events in the walking process to be carefully distinguished. One is a *step* to a higher fitness. The other is a *trial* of a one-mutant neighbor which may or may not have a higher fitness. This latter event may not lead to any movement away from the current position in sequence space. In our model, we equate steps to higher fitness with point mutations that lead to higher affinity antibodies (or more precisely immunoglobulin receptors). We assume that expression of these improved mutations causes B-cell clonal expansion that can be experimentally observed during an immune response. Trials which result in antibodies with lower fitness than that of the current antibody are assumed not to be expressed or to lead to cell lineages that do not expand significantly. Why expansion does not occur is not known, but the empirical facts appear to be that mutants with lower fitness than the germline antibodies are rarely if ever observed. For example, Manser[23] analyzed 20 clonally related mutants, 19 of which had roughly equal or higher affinity than the unmutated precursor.

Our model follows the sequential changes in a single antibody sequence. There are two possible ways in which this model can be related to biological reality. One possibility is to consider only those trials that are steps and hence result in a move to higher fitness. This view corresponds to analyzing only those lineages of accumulated mutations that are known to always lead in the direction of increasing affinity. In this scenario, we do not need to appeal to selection to pull a walk uphill. Alternatively, if we include in our consideration all attempted mutations, then we need a mechanism for allowing the process to stay in place until the first fitter mutant is found. For example, we can assume that at each cell division, at least one of the daughter cells is an unmutated copy of the parent cell. (Manser, this volume, suggests an alternative scheme). We must also appeal to selection to ensure that once a fitter mutant is tested, it will be expressed at a higher rate than the parent antibody configuration, effectively "swamping" the expression of the mutant with lower fitness. Clearly, it is unrealistic to imagine an instantaneous replacement of an antibody by its fitter one-mutant variant. Indeed, the relationship between B lymphocyte proliferation rate and receptor affinity for stimulating antigen is not known. However, as we show below, this simplistic model leads to insights into the somatic evolutionary process.

3. RESULTS

We focus on graphical representations of the predictions of our model, the mathematical analysis of the model having been published elsewhere.[22] Many of the key results have been proven to be independent of the particular selection of distribution of fitness values.[22] Whenever this distribution affects the·model's predictions, we assume that affinities have a lognormal distribution, truncated to lie within feasible bounds. This assumption is equivalent to assuming that the free energy of binding has a normal distribution. Besides the lognormal, immunologists have also used the Sips distribution[9,11,33] to describe antibody affinities. However, the Sips distribution is not well behaved—all of its moments are infinite[15]—and thus we will not use it here.

We denote antibody affinities by U. In our model, affinities are determined probabilistically from antibody sequences. Hence, we need to introduce some notation from probability theory. We denote the probability distribution $Pr(U \leq u)$ by $G(u)$. Thus $G(u)$ is the probability that an antibody has an affinity U that is less than or equal to u. Probability densities may be more familiar. Roughly speaking the *probability density* $g(u)$ is the probability that the affinity is u.[2] The formal relationship between the probability distribution $G(u)$ and its density $g(u)$ is given by

$$G(u) = \int_{-\infty}^{u} g(x)dx .$$

(1)

[2]More precisely, $g(u)du$ is the probability that the affinity is between u and $u + du$.

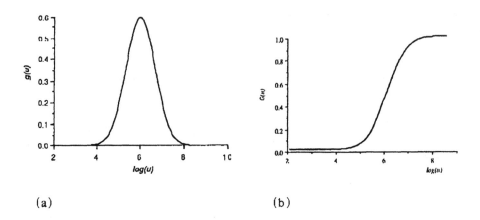

(a)   (b)

FIGURE 2 The lognormal distribution with parameters $\mu = 6$ and $\sigma = 2/3$. (a) The probability density $g(u)$ plotted versus $\log_{10}(u)$. (b) The cumulative distribution $G(u)$ versus $\log_{10}(u)$.

Some results given below will depend on fitness values. In these cases, we assume the affinity distribution is lognormal, truncated to lie between 10^2 and 10^{10}. This range is somewhat larger than is usually measured for serum antibodies. In fact, B cells with affinities less than 10^4 or 5×10^4 are probably not triggered by antigen during an immune response. In our model, antibodies with affinity $u < 10^4$ will be rare but are included for reasons of symmetry in the underlying free-energy distribution. Figure 2 illustrates the distribution we use. To characterize the distribution, we note that the logarithm of a lognormal random variable is distributed according to a normal distribution with mean μ and standard deviation σ. We have chosen $\mu = 6$ to ensure that the mean antibody affinity in the model is close to 10^6 (but not precisely 10^6 because the lognormal distribution is not symmetric and has a long tail at high affinities). The mean affinity that we use is close to that of the unmutated precursor in the experiments described by Manser.[23,24] Affinities higher than 10^8 are rarely seen during affinity maturation experiments (e.g., Manser[23]), and the value of σ that we chose, $\sigma = 2/3$, assigns a probability of less than 1% of attaining an affinity of 10^8 or greater purely by chance selection of an antibody configuration. The form of the truncated lognormal density function that we use is

$$g(x) = \frac{K}{x} \exp\left\{ -\frac{1}{2} \left(\frac{\log_{10} x - \mu}{\sigma} \right)^2 \right\} , \qquad (2)$$

where K is a constant chosen to ensure that $g(x)$ integrates to 1 over the allowable range of x values, $\mu = 6$ and $\sigma = 2/3$. For these values of μ and σ, the mean and standard deviation of a random variable with density g are 3.2×10^6 and 10^7, respectively. The density g is illustrated in Figure 2(a).

3.1 CHARACTERIZATION OF THE LANDSCAPE

Intuitively, we expect a rugged landscape to have many local optima. If a sequence has a fitness higher than all of its D one-mutant neighbors, it is a local optimum. Because fitnesses are chosen at random from the same probability distribution, the probability that a particular sequence has a fitness higher than D others is $1/(D+1)$. Thus the average number of local optima is equal to a^N, the number of possible sequences, multipled by $1/(D+1)$, the probability that a sequence is a local optimum. One can obtain more information about the number of local optima. If we denote the number of local optima by S_N, then, *for any distribution* G, as N becomes large,[3,21] the distribution of S_N approaches a normal distribution with mean $a^N/(D+1)$ and variance $a^N[D-(a-1)]/2(D+1)^2$. An accurate *ad hoc* argument for the independence of this result from the distribution of G is that, when sequences are independently assigned fitnesses, the only factor affecting progress toward a local optimum is the number of currently available one-mutant neighbors that have higher fitnesses. Thus, the complete information on progress of a walk is contained in the rank of the fitnesses. Equiprobable rank-ordering of fitnesses can be induced by sampling independently according to any distribution.

If local optima occurred at random throughout the landscape, the asymptotic distribution for S_N would be Poisson, for which the mean and variance are equal. However, our model predicts that, for $N \rightarrow \infty$, $var(S_N)/E(S_N) \rightarrow \frac{1}{2}$. The fact that local optima cannot be adjacent in sequence space induces a weak correlation among points in sequence space, leading to less variability than expected under the random model, and a different asymptotic distribution (i.e., normal, rather than Poisson) for S_N.

It is straightforward to extend the calculation of the asymptotic distribution for S_N to allow either one or two mutations to occur per cell generation. In this case we say a sequence is a local optimum if it has a fitness higher than that of all 1- and 2-mutant neighbors. Such optima are rarer than in the case considered above where an optimum has only to be fitter than its 1-mutant neighbors. In fact, the mean of the limiting distribution of S_N, which is still normal, is $a^N/[D+1+D(D-1)/2]$.[16] Because there are fewer optima, walks can be expected to be rather longer than in situations in which only one mutation can be performed per generation.

3.2 CHARACTERIZATION OF WALKS

There are two essential features of walks on a rugged landscape that we will quantify using our model. One is the length of walks to a local optimum; the other is the fitness attained at positions along a walk. In the following definitions and subsequent analyses, the term *step* will be used exclusively to describe mutations leading to an increase in fitness. Steps should be distinguished from *trials of mutational variants*, which may or may not lead to an increase in affinity of antibody.

To describe the lengths of walks to an optimum, we measure three quantities. For walks beginning at a fitness u_0, we first calculate $W(u_0)$, the total number of *steps* to reach a local optimum. Since in our model, steps correspond to mutations leading to a higher fitness, the length of a walk may be compared with experimentally observed numbers of expressed mutations in primary and secondary antibodies. Second, we calculate $M(u_0)$, the total number of *distinct mutations* tested along a mutational walk. The third variable of interest is $T(u_0)$, the total number of *not necessarily distinct* mutations which are tested on a mutational walk. By comparing the number of distinct mutations tested with the total number of mutations tested, we gain some insight into the efficiency of the mutational process. An efficient process would be expected not to retry a mutation leading to an inferior antibody.

In addition, we measure two quantities that describe in more detail progress toward an optimum *at a particular stage* along a path. For paths that have reached a fitness u, these quantities are $m(u)$, the number of distinct mutations tested at fitness u, before finding a fitter variant; and $t(u)$, the number of (not necessarily distinct) mutations tested at fitness u, before finding a fitter variant. If we assume one attempted mutation occurs per cell generation, we have a measure of the rate of expression of antibody with increasing affinities. We might expect, and in fact find to be true, that as u increases, an increasing number of variants must be tested before the first fitter one is found, i.e., the affinity maturation process slows down. As u approaches very high values, the process slows so much that it might be difficult to distinguish experimentally between reaching a local optimum and reaching a high fitness value where progress stagnates. We also find that the proportion of repeated mutants is very small unless a path reaches a very high fitness.

The other essential feature of a walk is the fitness attained along a path. If we examined many independent walks starting from the same fitness u_0, we would find that they explored very different routes across the landscape, reaching different sequences and fitnesses at each step along their path starting at fitness u_0. To describe the variety of possibilities, we calculate the probability density functions that describe the random nature of fitnesses attained along a path. Three useful density functions are: $f_k(u; u_0)$, the density of fitnesses attained on the kth step of an evolutionary walk; $f_{\text{evol}}(u; u_0)$, the density of fitnesses attained at the end of a walk to a local optimum; and $f_{\text{rand}}(u)$, the density of fitnesses attained by a random sampling of sequences. Using $f_k(u; u_0)$, we can examine predictions of the model about, for example, long walks. We might ask, does the model predict that long walks are more likely to reach unusually high fitnesses or to simply muddle around at inferior fitnesses? A comparison of the densities f_{evol} and f_{rand} will give another type of insight into the efficiency of the mutational process. The expectation would be that an evolutionary process should reach higher affinities than simply choosing antibody sequences at random according to some reasonable scheme.

In the following subsections, we examine each of these random variables and densities in turn. In addition to presenting analytical results, we evaluate the formulae for typical values of D and u_0. Graphs are presented for $D = 1500$; all calculations for D of this magnitude are quite insensitive to variations in D of at least $\pm 10\%$.

The starting point for walks, u_0, was chosen to be 10^6, for which $G(u_0) = 0.5$. That is, 10^6 is the median fitness from our lognormal distribution (see Figure 2(b)). Since the lognormal distribution is highly skewed with a long upper tail when plotted versus u rather than $\log u$, the mean fitness, 3×10^6 is higher than the median, and in fact is the upper 76 percentile of the fitness distribution. We chose to start at the median fitness rather than the mean, so as to represent a middle case. The results that follow are not very sensitive to choices of the starting fitness below the median. If one chose $u_0 = 10^5$ or 5×10^4 rather than 10^6, results would change little. The reason for the insensitivity to u_0 is that if one started at the lowest possible fitness, 10^2, then, on average, after the first uphill step an antibody would have a fitness half-way to the top, i.e., 10^6. Thus, the average number of steps to an optimum, computed below, will differ by at most one step as one varies the starting fitness between 10^2 and 10^6.

3.2.1 $W(u_0)$, NUMBER OF STEPS TO AN OPTIMUM

We have previously proven that the distribution of $W(u_0)$ is independent of the particular probability distribution of fitness values assumed in the model.[22] The probability $p_W(k; u_0)$ that a walk to a local optimum, starting at fitness u_0, is k steps long is

$$p_W(k; u_0) = G^D(u_0) , \quad k = 0 , \tag{3a}$$

and

$$p_W(k; u_0) = \frac{1 - G^D(u_0)}{1 - G(u_0)} \frac{1}{(k-1)!} \int_{G(u_0)}^{1} [V(x) - V(G(u_0))]^{k-1} x^{D-1} dx , \tag{3b}$$
$$k \geq 1 ,$$

where

$$V(x) = \int_0^x \frac{1 - z^{D-1}}{1 - z} \, dz . \tag{4}$$

If D is very large and u_0 is not too close to the global optimum so that $D(1 - G(u_0)) \gg 1$, $p_W(k; u_0)$ can be approximated by

$$p_W(k; u_0) \sim \frac{1}{(k-1)!} \frac{1}{D(1 - G(u_0))} \int_0^\infty \left(\ln \left[\frac{D(1 - G(u_0))}{x} \right] - E_1(x) \right)^{k-1} e^{-x} dx ,$$
$$k \geq 1 , \tag{5}$$

where E_1 is the exponential integral.[1] For antibodies, $D = 1500$ is sufficiently large that the condition $D(1 - G(u_0)) \gg 1$ holds for almost all starting fitnesses. In fact, for this condition to be violated, one would need to start at a fitness in about the upper 99.93 percentile. We present the probability distribution of $W(u_0)$ for $D = 1500$ and $u_0 = 10^6$ in Figure 3.

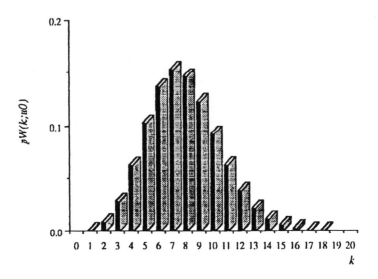

FIGURE 3 The probability $p_W(k; u_0)$ that a walk to a local optimum, starting at affinity $u_0 = 10^6$, ends on the kth step.

The condition $D(1-G(u)) \gg 1$ will appear in many of our approximate results. The condition defines a "boundary layer" near the global optimum $G(u) = 1$, where the mathematical behavior of the model changes character. A mathematical derivation of this fact is given in Macken, Hagan and Perelson.[22] The boundary layer arises from examining the probability of being at a local optimum. From Eq. (3a) we see that at the starting fitness u_0, this probability is just $G^D(u_0)$. That is, u_0 is a local optimum if all of its D neighbors have lower fitness. The probability of a neighbor having lower fitness is $G(u_0)$ and thus the probability of all D neighbors having lower fitness is $G^D(u_0)$. At a fitness other than the starting fitness, the probability of a local optimum is $G^{D-1}(u)$. One need examine only $D-1$ neighbors since it is known that one neighbor, the one the walk came from, has lower fitness. To see where the boundary layer comes from, note that for large D,

$$G^{D-1} \approx G^D = \exp[\ln(G^D)] = \exp[D \ln(1 - (1 - G))] .$$

Expanding the logarithm in a power series, one finds

$$G^D = \exp[-D(1 - G) - D(1 - G)^2/2 - \ldots] .$$

If $D(1 - G(u)) \gg 1$, then G^D is essentially zero (i.e., transcendentally small); it is only when $D(1 - G(u))$ becomes of order one that this probability differs substantially from zero. From the existence of this boundary layer we can conclude many things. For example, we conclude that almost all local optima have fitnesses near that of the global optimum.

From the distribution for $W(u_0)$, Eq. (3), we obtain the moments of $W(u_0)$. For large D the mean, $E[W(u_0)] \sim 1.0991 + \ln[D(1 - G(u_0))]$ and variance $var[W(u_0)] \sim 0.26 + \ln[D(1 - G(u_0))]$. Notice that the average walk length increases slowly with D. For the parameter values $\mu = 6$, $\sigma = 2/3$, $u_0 = 10^6$, the mean walk length is 7.7 steps with a standard deviation of 2.6 steps. Less than 9% of the walks will be shorter than 4 steps or longer than 12 steps. If we had chosen $u_0 = 4.8 \times 10^6$, so that $G(u_0) = .8$, then the mean walk length would be 6.8 steps with a standard deviation of 2.4 steps. The results are robust to the selection of u_0, as long as u_0 is chosen outside the boundary layer. This robustness is a reflection of the fact that walks starting outside the boundary layer move to higher fitnesses in a very few steps. In the chapter by Kauffman and Weinberger, in which the NK model is used to represent the landscape, walks are started in the boundary layer. This is an important distinction between the application of these different models to affinity maturation that we will return to in Section 4.

3.2.2 $M(u_0)$, TOTAL NUMBER OF DISTINCT MUTATIONS

As is the case for the random variable $W(u_0)$, the random total number of distinct mutations tested along a path to a local optimum, $M(u_0)$, has a distribution that is independent of the particular distribution of fitness values, G, assumed in the model. For large values of D, the mean $E[M(u_0)] \sim 0.781D$, with standard deviation $\sim 0.624D$, provided $D(1 - G(u_0)) \gg 1$. That is, provided we do not start at a fitness u_0 in the upper 99.93 percentile of fitnesses, the distribution of $M(u_0)$ is independent of u_0. Thus, for starting fitnesses outside the boundary layer, and with $D = 1500$, the average number of distinct mutations tested on a path to a local optimum is 1172, with a corresponding standard deviation of 936. Notice that the process is extremely variable.

In Section 3.2.1 we predicted that the mean number of steps to an optimum is 7.7. Thus we predict that on average, 7.7/1172 or approximately 0.66% of mutations will be improvements.

By focusing on distinct mutations, we assess the proportion of sequence space explored by a mutational process proceeding according to our rules. Sequence space contains $20^{230} \simeq 10^{300}$ antibody V-region sequences, of which only 1172 are examined. Thus, the percentage of possibilities explored by the mutational process is extremely small, around $10^{-295}\%$.

3.2.3 $T(u_0)$, TOTAL NUMBER OF NOT NECESSARILY DISTINCT MUTATIONS

As in the cases for $W(u_0)$ and $M(u_0)$, the distribution of $T(u_0)$ does not depend on the fitness distribution G. The random variable $T(u_0)$ roughly reflects the duration of an evolutionary walk from a starting fitness u_0 until it reaches a local optimum. The strength of correspondence of $T(u_0)$ with time depends on the relationship between the occurrence of mutations and cell divisions, and the effect on cell cycle time of changes in receptor affinity for antigen. Manser (this volume) calls into question the hypothesis that mutation and cell division are coupled, so that associating $T(u_0)$ with time may not be straightforward.

In Macken, Hagan and Perelson[22] we calculate that for large D, $E[T(u_0)] \sim 1.224D$, for any u_0 such that $D(1 - G(u_0)) \gg 1$. For $D = 1500$, this mean is 1836. Of the total number of trials taken to reach an optimum, on average $\{T(u_0) - M(u_0)\} \sim .44D$, or approximately 36% of all trials, will be repeats. Most of these repeats occur when the fitness is close to a local optimum and many trials are needed to find a fitter variant. A comparison of $m(u)$ and $t(u)$ in the sections below will enlarge on this aspect of the mutational process.

3.2.4 $m(u)$, NUMBER OF DISTINCT MUTATIONS TESTED AT FITNESS u Our model assumes that at each step of the mutational walk, the D one-mutant neighbors are tested in random order until the first one with a higher fitness than the currently expressed mutant is found. At a given fitness $u > u_0$, the testing process may be "lucky," in that a fitter mutant is found quickly. On the other hand, the search may continue through many trials before a fitter mutant is found. We also may expect that on average the process will take longer to improve at higher values of u than at lower values. To consider these issues, we have computed the probability

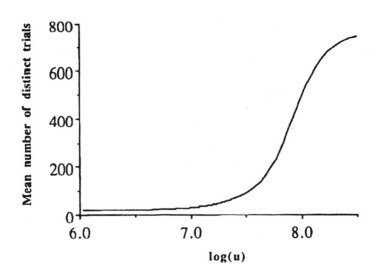

FIGURE 4   The mean number of distinct trials at fitness u, plotted against $\log_{10}(u)$, where we have assumed that u is not a local optimum.

distribution of the number of trials to leave a fitness u, given that u is not a local optimum. In Macken, Hagan and Perelson[22] we show that

$$p_m(k; u) = Pr\{m(u) = k|u \text{ is not a local optimum}\}$$
$$= \frac{1 - G(u)}{1 - G^{D-1}(u)} G^{k-1}(u), \quad k = 1, 2, \ldots, D - 1. \tag{6}$$

When $D(1 - G(u)) \gg 1$, the mean and variance of this distribution are asymptotically $1/(1 - G)$ and $G/(1 - G)^2$, respectively. Figure 4 exhibits the conditional mean of $m(u)$. Around affinities of 3×10^7, we observe a sudden increase in the number of trials. The path is now at a fitness approximately in the upper 99 percentile of the distribution, where it is much harder to find fitter variants of the current mutant. In fact, it is quite likely that sequences with fitnesses in this percentile are local optima (see Section 3.2.7).

3.2.5 $t(u)$, NUMBER OF NOT NECESSARILY DISTINCT MUTATIONS TESTED AT FITNESS u

Suppose that an evolutionary walk has achieved a fitness u greater than the starting value u_0. Now, one-mutant neighbors of the current mutational variant are tested in random order until the first fitter mutant is found. It is hard to imagine that the affinity maturation process has any built-in mechanism to ensure that the same mutation is not tested more than once. In our model, as long as it is easy to find a one-mutant neighbor with a higher fitness, it is unlikely that the mutational process will repeat a trial. However, as the number of samples is increased, the probability of sampling the same one-mutant neighbor more than once purely by chance increases. In fact, for n trials out of D equally likely possibilities, the probability that no trial is repeated is $1(1 - 1/D)(1 - 2/D) \ldots (1 - (n - 1)/D)$, clearly close to 1 unless n is nearly as large as D. For paths that have not reached the boundary layer, the expected value of $t(u)$ given that u is not an optimum is $1/(1 - G)$, which is the same as the conditional mean of the number of distinct trials, $m(u)$. For fitnesses very close to a local optimum (i.e., in the boundary layer), the means of $t(u)$ and $m(u)$ diverge, as shown in Figure 5. The conclusion from Figure 5 is that at fitnesses around 5×10^7, mutants will start to be retested. At $u = 5 \times 10^7$, $G(u) = .994$. Thus, paths reach the upper 99.4 percentile of fitnesses with little retesting.

Since we guess that the mutational process observed experimentally is operating in the region in which $D(1-G) \gg 1$, i.e., outside the boundary layer around $G(u) = 1$, we expect that most of the tested mutants will be distinct. The mutational process is therefore likely to be efficient in this sense.

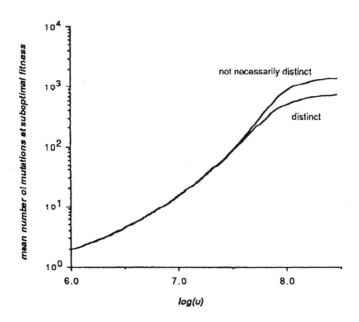

FIGURE 5 The mean number of mutations tested at affinity u, given that u is not a local optimum. For affinities below 5×10^7 very few mutations are retried and thus the process is very efficient. At affinities above 5×10^7 the number of possible improvement mutations has decreased significantly and it becomes difficult to find new ways uphill.

3.2.6 $f_k(u; u_0)$, DENSITY OF FITNESSES ATTAINED ON THE kth STEP OF A WALK HAVING AT LEAST k STEPS Paths which continue for many steps without reaching a local optimum could be exhibiting one of two quite different behaviors: they continue for a long time because they take many steps at low fitnesses, far from a local optimum; or they are one of the rare paths which attains a high fitness without first becoming trapped at a local optimum. We gain some insight into which of these two scenarios is more likely by examining the fitness attained on the kth step of paths having at least k steps, for $k = 1, 2 \ldots$. The fitness attained is a random variable, and so we describe it by its probability density function, mean and variance. For paths starting at u_0, we calculate the density function[22] as

$$f_k(u; u_0) = \frac{1 - G^D(u_0)}{1 - G(u_0)} \frac{g(u)}{(k-1)!} \left[V(G(u)) - V(G(u_0)) \right]^{k-1}, \quad k \geq 1, \quad (7)$$

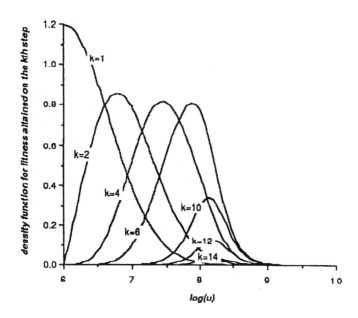

FIGURE 6 The probability that fitness u is attained on the kth step of a walk of at least k steps starting at affinity $u_0 = 10^6$.

where $V(G)$ is given in Eq. (4) and $g(u)$ is the lognormal density function from Eq. (2).

The density function f_k clearly involves the distribution of fitnesses as well as the starting fitness u_0. The graphs of f_k in Figure 6 show a family of curves with the mode and mean shifting progressively toward higher affinities as the number of steps, k, increases. Figure 6 also shows the area under the curves decreasing as k increases. For a particular value of k the area represents the probability that a path continues for at least k steps. Since, as k increases, the chance of reaching a local optimum on or before the kth step increases, we see the reason for the "shrinking" of the f_k curves.

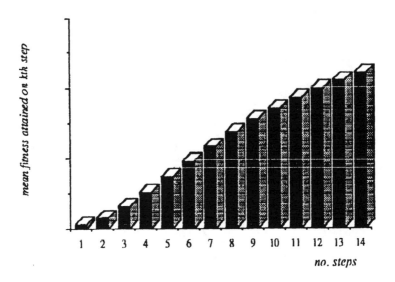

FIGURE 7   The mean fitness attained on the kth step of a walk of at least k steps starting at fitness 10^6.

To examine the apparent shifting of the mean of the curves more carefully, we condition on a path having at least k steps, and then calculate the mean and standard deviation of the fitness attained on the kth step. Hence, we define $U_k(u_0)$ to be the fitness attained on the kth step for a path starting from fitness u_0, and calculate[22]

$$E[U_k^n(u_0)|\text{path has at least k steps}] =$$
$$\frac{\int_{u_0}^1 u^n \ [V(G(u)) - V(G(u_0))]^{k-1} \ g(u) \ du}{\int_{G(u_0)}^1 [V(x) - V(G(u_0))]^{k-1} \ dx} \ , \quad k \geq 1; \ n = 1, 2 . \quad (8)$$

The mean fitness is shown in Figure 7. For paths with at least six steps that start at fitness 10^6, the average fitness attained on the sixth step is 9.6×10^7 with a standard deviation of 1.09×10^8. If a path continues for as many as 14 or more steps, then the average fitness attained on the 14th step is 2.2×10^8 with a standard deviation of 1.8×10^8. It is informative to observe that the coefficient of variation (standard deviation/mean) of the fitness attained decreases as k increases.

To summarize, we find that it takes rather few steps to improve substantially in fitness. Here an increase of two orders of magnitude occurs in six steps. If paths continue for many steps, they do so virtue of muddling around at high fitnesses where little further progress is made (the next eight mutations only improve fitness by an extra factor of two).

3.2.7 $f_{\text{evol}}(u; u_0)$, DENSITY OF FITNESSES ATTAINED AT THE END OF A WALK   Two walks starting at identical positions in sequence space, with the same fitness u_0, will almost certainly reach different local optima. The reason for the high probability of divergence of paths is the high dimensionality of sequence space, leading to an extremely large number of options for directions of movement along a path. In fact, even if several paths are initiated from the same position in sequence space, it is probable that each walk will end at a different local optimum. Thus, the fitness

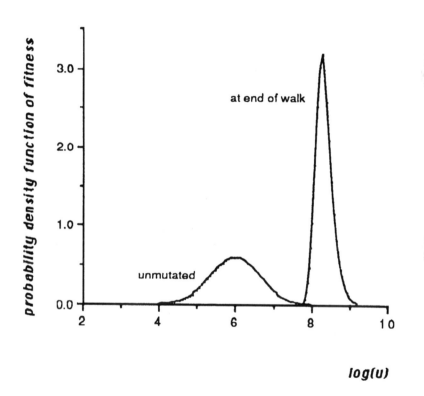

FIGURE 8   A comparison of the fitness attained at the end of an evolutionary walk ($f_{\text{evol}}(u; u_0)$, given by Eq. (9)) with a fitness chosen at random from the lognormal distribution of antibody affinities ($g(u)$, given by Eq. (2)). The evolutionary walk was assumed to start at fitness 10^6 and to continue until a local optimum was reached. The graph demonstrates the efficiency of evolutionary walks in picking out high-affinity antibodies.

attained at the end of an individual path, which we shall denote by $U_{evol}(u_0)$, is unpredictable, but we can describe the likelihood of different values by the probability density function of U_{evol}, namely $f_{evol}(u; u_0)$.

For walks starting outside the boundary layer, we have shown that[22]

$$f_{evol}(u; u_0) \sim \frac{g(u)}{1 - G(u)} \exp \left\{ - D(1 - G(u)) - E_1[D(1 - G(u))] \right\} . \tag{9}$$

We can use f_{evol} in two ways to study the efficiency of the mutational process. First, a simple relevant measure of the success of a mutational walk is the average rank of the fitness finally attained. In Macken, Hagan and Perelson,[22] we show that this average, after normalizing so that the rank lies between 0 and 1, is $1 - .6243/D$, or 99.958% (for $D = 1500$), with a standard deviation of $0.68/D$. The average rank corresponds to a fitness of 1.7×10^8. Second, we can compare the fitnesses attained at the end of walks with the fitnesses attained if the antibody were simply chosen at random from all possible antibodies. Figure 8 compares $f_{evol}(u; u_0)$, the probability density function for the affinity attained at the end of a walk with $g(u)$, the probability density function for the affinities of antibodies in sequence space. Of course, if more than one random sample from g was allowed, then the maximum of all values sampled would probably be higher than the affinity of just a single sample. In the next section, we will compare U_{evol} with the maximum of several independent samples from g.

3.2.8 $f_{rand}(u; u_0)$, DENSITY ATTAINED BY RANDOM SAMPLING FROM G Suppose r random selections are made from sequence space. We will denote the maximum affinity of these r random selections by $U_{rand}^{(r)}$. By the arguments of Section 3.2.1, we calculate that $G^r(u)$ is the probability that $U_{rand}^{(r)} \leq u$. From this, we obtain the probability density function of $U_{rand}^{(r)}$, namely

$$f_{rand}(u) = rG^{r-1}(u)g(u) . \tag{10}$$

The average normalized rank of $U_{rand}^{(r)}$ is $1 - 1/(r + 1)$, which is smaller than the average normalized rank of U_{evol} for all values of r less than 2402. Thus, unless at least 2402 different sequences are sampled, the mutational walk will reach a sequence with a higher fitness on average. We saw in Section 3.2.2 that the average number of distinct trials performed on a mutational walk is 1172, starting from $u_0 = 10^6$. Even if we had started a walk from the lowest possible fitness, the average number of distinct trials is still 1172. The mutational walk is therefore more efficient than random sampling from the space of all possible sequences.

4. DISCUSSION

We have shown that a simple model in which antibodies are assumed to evolve by point mutations that randomly change the affinity of the antibody for the immunizing antigen, coupled with selection for higher affinity variants, can explain much of the quantitative and qualitative data on affinity maturation. For example, one of the puzzling features of the somatic mutation process is that it generally only leads to one or two orders of magnitude improvement in antibody affinity, after which further mutations lead to little or no improvement. In experiments by Berek and Milstein[4,5] when further improvements are seen they generally are due to the evolution of a new germline sequence rather than further improvements by point mutation. Both of these features are predicted by our model. The halting in affinity improvement is predicted to occur due to trapping at a local optimum or near a local optimum where improvement mutations become rare (Figure 5). Different germline sequences correspond to different starting positions on the landscape. Thus different germline genes should get trapped at different peaks, and thus one might see in secondary and tertiary responses the attainment of high affinities being achieved by rare germline V-region gene combinations.

In addition to capturing these qualitative features of the somatic mutation process, the model makes a number of quantitative predictions. It predicts that 7.7 ± 5.2 replacement mutations (mean $\pm$ 2 s.d.) should occur before an antibody stops improving in affinity. Interestingly, this is roughly the range of variation seen in experiment. According to the model, these 7.7 ± 5.2 replacement mutations leading to higher affinities constitute approximately 0.7% of the mutant antibodies tested along an evolutionary walk. Experiments have not yet determined the fraction of mutations that lead to higher affinity antibodies, but this prediction is clearly testable.

The model also predicts that the first few improvement mutations substantially increase the affinity and that later mutations only make small improvements. From Figure 7 one can see that an antibody starting at an affinity of 10^6 is predicted to reach affinities of 5×10^7 after 4 steps, 10^8 after 6 steps, and 2×10^8 after 12 steps. Because walks have only 7.7 steps on average, with a standard deviation of 2.6 steps, it is unlikely for a walk to continue for as long as 12 steps and reach an affinity of 2×10^8. In fact, only about 6% of walks should do so. The model also makes a testable prediction about the fraction of mutations of the unmutated precursor antibody that are improvements. For a random landscape, starting at the 50th percentile implies that 50% of one-mutant variants of the starting antibody should be fitter. Starting at the 70th percentile would imply that 30% of one-mutant variants are fitter, and so forth. To obtain a sensible prediction about the number of steps to a local optima, a walk must start far outside the boundary layer, say, at the 90th percentile or lower. Thus measuring the relative affinities of the one-mutant neighbors of the starting antibody should provide a strong test of the random fitness model.

It seems surprising that a model in which fitnesses are assigned at random can lead to reasonable predictions. One means of rationalizing this result is that if affinities are chosen from a lognormal distribution with a small half-width, as in Figure 2, then the affinities of an antibody and its one-mutant neighbors will most-likely be similar. Thus, even though there is no correlation present in the model one will generally only observe small changes in affinity. Because, the model is random these changes can be either an increase or decrease in affinity. Further, on rare occasions, at frequencies determined by the variance in the distribution, a mutant will be chosen that has a much higher or much lower affinity than average. This is typically what is observed.

Comparing with the chapter by Kauffman and Weinberger (this volume), one sees that the NK model makes similar predictions. Thus different landscape models can give rise to similar predictions. There are three major differences between the application of the NK model and the random landscape model used here. First, Kauffman and Weinberger assume that the starting point of the walk is at a fitness in the 99.999 percentile. This corresponds to the fact that 1 out of 10^5 B cells typically respond to a given epitope and thus one might assume that the starting antibody is the fittest antibody out of a sample of 10^5 antibodies. We, on the other hand, assume that the only B cells under consideration are those that respond to antigen. Thus we choose our range of affinities to correspond with those seen in a typical somatic mutation experiment. Further, we choose our initial condition at the 50th percentile, so that the starting configuration is a typical antibody directed at the immunizing antigen. As shown in Figure 8, we then find that an antibody starting at affinity 10^6 could attain an affinity of 10^8 before being trapped at a local optimum. However, it would be unlikely for an antibody to make it to the 99.999 percentile; in our model this would correspond to an affinity of 6.7×10^8. The starting affinity and the range of affinities that we predict for somatic variants are in rough accord with those seen by Manser (also Manser, this volume).[24]

A second difference between the models is the distribution of affinity values. Here we have used a lognormal distribution. In the NK model, N random fitness contributions are averaged to determine the fitness of an antibody. When a large number of random variables are averaged, their distribution will approach a normal distribution with a small variance, regardless of the parent distribution. Thus in the NK model fitnesses are essentially normally distributed rather than lognormally distributed. This distinction could easily be overcome by multiplying the fitnesses in the NK model rather than adding them.

A third difference between our model and the Kauffman-Weinberger model is that we model the antibody combining site as being composed of both heavy- and light-chain variable regions, and thus having approximately 230 amino acids. Kauffman and Weinberger consider only a single variable region of 112 amino acids. For this single variable region they estimate that $K = 40$, i.e., that 40 amino acids influence the fitness of each amino acid in the variable region. Although this value is high, it is plausible. Examining their Table 1b, it seems as if K would need to increase substantially if N were doubled, so as to maintain the mean walk length around 8. From their Table one might speculate that K might have to be 80 or larger.

Further work will be needed to decide between the random landscape model we described here and the NK model with $K = 40$. Most likely, neither is correct and refinements will lead the way to better models. It is difficult to believe that a model in which affinities are assigned to sequences totally randomly can be correct in detail. There are however means of adding correlations to a model that will still preserve the lognormal distribution of affinities and allow analytical predictions to be made. For example, J. Felsenstein, University of Washington, suggested a model in which the antibody sequence is broken up into a set of independent domains, with each domain making a random fitness contribution u_i, and the total fitness u chosen as the product of the u_i. A lognormal distribution of u values will result if the number of domains is large. Further, because a single mutation will at most affect the fitness of a single domain, the fitnesses of one-mutants neighbors will be correlated. This model is similar to the NK model, but because it uses independent domains rather than domains of size K chosen at random, it can be treated analytically.

An alternative model of fitness is a variant of the now-classical additive fitness model, in which the fitness of the immunoglobulin is determined by the average of the fitnesses of the amino acid at each position. Provided the fitness of an amino acid at one position is assigned independently of the configuration of the rest of the immunoglobulin, a random mutation process that leads to a strictly uphill walk on the landscape can be studied analytically in detail. Some results have been presented previously in the context of an NK model[17] with $K = 0$ where it was shown that one obtains a fitness landscape that is highly correlated. The reason that the landscape is correlated is that changing one amino acid changes the fitness of the entire molecule by at most $1/N$, where N is the length of sequence. The landscape has only one optimum that corresponds to the antibody with the fittest amino acid at each position. For an antibody with $N = 230$ amino acids in the variable region and any of the twenty amino acids allowed in each position $(a = 20)$, we have computed that the mean number of steps to the optimum, starting with all amino acids at their median fitness, is 673.7 with a standard deviation of 17.8 (Macken and Perelson, in preparation). This prediction clearly does not correspond to observations on antibody evolution.

A natural next step is to analyze cases intermediate between the random uncorrelated landscape and the additive highly correlated landscape. The NK model and the domain model suggested by Felsenstein both are means of doing this. Another way is to assume that the positions in a protein can be classified into, say, three types. Changes at one type of position greatly affect the fitness of the molecule and can be modeled as changing the fitness randomly. Changes at the second type of position (e.g., far from the combining site of the antibody) change the fitness in a minor way and thus can be modeled as a correlated change. The third type of position might be one that has been evolutionarily conserved, such that a change in that position leads to a molecule with zero or greatly reduced fitness. Proteins then fall into classes, according to the fraction of positions of each type. Proteins within a given class have a characteristic landscape. Changing the proportions of each type of position changes the ruggedness and other characteristics of the landscape. In the case of antibodies, framework and complementarity determining regions (CDRs) have already been identified. The CDRs are known to be very variable and thus almost any substitution should be allowed. This is not the case in the framework, where some positions are known to be evolutionarily conserved. Thus as an initial examination of an intermediate model one might assume that substitutions in the CDRs change affinity in a random way, that substitutions in the evolutionarily conserved framework positions reduce the fitness of the molecule, and that substitutions at the remaining positions in the framework affect the affinity in a minor way. Labeling positions by type requires information about structure and the structural consequences of point mutations.

The immune system has been a testing ground for many ideas in molecular biology. Affinity maturation by somatic mutation seems an ideal system in which nature is allowing us to examine in detail the relationships between antibody sequence, antibody structure, and antibody function. With appropriate models one may not only be able to infer information about antibody structure from measurements of affinity and sequence, but also get glimpses of the evolutionary driving forces and operating principles of importance in the immune system.

ACKNOWLEDGMENTS

This work was done under the auspices of the U.S. Department of Energy. C. A. Macken would like to thank the Santa Fe Institute for hosting her visits during which time much of this work was done. A. S. Perelson would also like to thank the Santa Fe Institute for their generous support of a Theoretical Immunology Program and the sponsorship of the workshop in which this paper was presented.

REFERENCES

1. Abromowitz, M., and I. A. Stegun. *Handbook of Mathematical Functions.* National Bureau of Standards, Washington, D.C., 1964.
2. Agur, Z., and M. Kerszberg. "The Emergence of Phenotypic Novelties Through Progressive Genetic Change." *Am. Nat.* **129** (1987):862–875.
3. Baldi, P., and Y. Rinott. "Asymptotic Normality of Some Graph Related Statistics." *J. Appl. Prob.* **26** (1989):171–175.
4. Berek, C., and C. Milstein. "Mutation Drift and Repertoire Shift in the Maturation of the Immune Response." *Immunol. Rev.* **96** (1987):23–41.

5. Berek, C., G. M. Griffiths, and C. Milstein. "Molecular Events During the Maturation of the Immune Response to Oxazolone." *Nature* **316** (1985):412–418.

6. Bruggeman, M., H.-J. Muller, C. Burger and K. Rajewsky. "Idiotype Selection of an Antibody Mutant with Changed Hapten Binding Specificity, Resulting from a Point Mutation in Position 50 of the Heavy Chain." *EMBO J.* **5** (1986):1561–1570.

7. Burnet, F. M. *The Clonal Selection Theory of Acquired Immunity.* Nashville, TN: Vanderbuilt Univ. Press, 1959.

8. Chien, N. C., V. A. Roberts, A. M. Giusti, M. D. Scharff and E. D. Getzoff. "Significant Structural and Functional Change of an Antigen-Binding Site by a Distant Amino Acid Substitution: Proposal of a Structural Mechanism." *Proc. Natl. Acad. Sci.* **86** (1989):5532–5536.

9. DeLisi, C. *Antigen Antibody Interactions.* Lect. Notes in Biomath., vol. 8. New York: Springer-Verlag, 1976.

10. Eigen, M. In *Emerging Syntheses in Science, Proceedings of the Founding Workshops of the Santa Fe Institute,* edited by D. Pines. Santa Fe Institute Studies in the Sciences of Complexity, Proc. Vol. I. Reading, MA: Addison-Wesley, 1988, 21–42.

11. Eisen, H. *Immunology,* second edition. Hagerstown, MD: Harper and Row, 1980.

12. Fontana, W., W. Schnabl, and P. Schuster. "Physical Aspects of Evolutionary Optimization and Adaptation." *Phys. Rev. A* **40** (1989):3301–3321.

13. Fleming, M., S. Fish, J. Sharon and T. Manser. "Changes in Epitope Structure Directly Affect the Clonal Selection of B Cells Expressing Germline and Somatically Mutated Forms of an Antibody Variable Region." Submitted for publication, 1990.

14. Gillespie, J. H. Molecular Evolution Over the Mutational Landscape." *Evolution* **38** (1984):1116–1129.

15. Goldstein, B. "Theory of Hapten Binding to IgM: The Question of Repulsive Interactions Between Binding Sites." *Biophysical Chem.* **3** (1975):363–367.

16. Kauffman, S. A., and S. Levin. "Towards a General Theory of Adaptive Walks on Rugged Landscapes." *J. Theoret. Biol.* **128** (1987):11–45.

17. Kauffman, S. A., E. D. Weinberger, and A. S. Perelson. "Maturation of the Immune Response Via Adaptive Walks on Affinity Landscapes." In *Theoretical Immunology, Part One,* edited by A. S. Perelson. Santa Fe Institute Studies in the Sciences of Complexity, Proc. Vol. II. Reading, MA: Addison-Wesley, 1988, 349–38.

18. Kauffman, S. A. and E. D. Weinberger. "The NK Model of Rugged Fitness Landscapes and its Application to Maturation of the Immune Response." *J. Theoret. Biol.* **141** (1989):211–245.

19. Kauffman, S. A. *Origins of Order: Self Organization and Selection in Evolution.* Oxford: Oxford Univ. Press, 1990.

20. Kocks, C., and K. Rajewsky. "Stepwise Intraclonal Maturation of Antibody Affinity Through Somatic Hypermutation." *Proc. Natl. Acad. Sci. USA* **85** (1988):8206–8210.

21. Macken, C., and A. S. Perelson. "Protein Evolution on Rugged Landscapes." *Proc. Natl. Acad. Sci. USA* **86** (1989):6191–6195.

22. Macken, C. A., P. Hagan, and A. S. Perelson. "Evolutionary Walks on Rugged Landscapes." *SIAM J. Appl. Math.*, 1990, in press.

23. Manser, T. "Evolution of Antibody Structure During the Immune Response. The Differentiative Potential of a Single B Lymphocyte." *J. Exp. Med.* **170** (1989):1211–1230.

24. Manser, T. This volume.

25. Maynard-Smith, J. "Natural Selection and the Concept of a Protein Space." *Nature* **225** (1970):563–564.

26. McKean, D., Huppi, K., Bell, M., Standt, L., Gerhard, W. and Weigert, M. *Proc. Natl. Acad. Sci. USA* **81** (1984):3180–3184.

27. Nino, J. *Approaches Moleculaires de l'Evolution.* New York: Masson, 1979.

28. Panka, D. J., M. Mudgett-Hunter, D. R. Parks, L. L. Peterson, L. A. Herzenberg, E. Haber and M. N. Margolies. "Variable Region Framework Differences Result in Decreased or Increased Affinity of Variant Anti-Digoxin Antibodies." *Proc. Natl. Acad. Sci. USA* **85** (1988):3080–3084.

29. Roberts, S., J. C. Cheetham, and A. R. Rees. "Generation of an Antibody with Enhanced Affinity and Specificity for its Antigen by Protein Engineering." *Nature* **328** (1987):731–734.

30. Schuster, P. "Potential Functions and Molecular Evolution." In *Chemical to Biological Organization*, edited by M. Markus, S. C. Mueller and G. Nicolis, 149–165. Berlin: Springer-Verlag, 1988.

31. Schuster, P. "Optimization and Complexity in Molecular Biology and Physics." In *Optimal Structures in Heterogeneous Reaction Systems*, edited by P. J. Plath. Series in Synergetics. New York, NY: Springer-Verlag, 1989, 101–122.

32. Sharon, J., M. L. Gefter, L. J. Wyoski and M. N. Margolies. "Recurrent Somatic Mutations in Mouse Antibodies to P-Azophenylarsonate Increase Affinity for Hapten." *J. Immunol.* **142** (1989):596–601.

33. Sips, R. "On The Structure of a Catalyst Surface." *J. Chem. Phys.* **16** (1948):490–495.

34. Weinberger, E. D. "A Rigorous Derivation of Some Properties of Uncorrelated Fitness Landscapes." *J. Theoret. Biol.* **134** (1988):125–129.

Tim Manser
Department of Biology, Princeton University, Princeton, NJ 08544

Maturation of the Humoral Immune Response: A Neo-Darwinian Process?

INTRODUCTION

During the course of a humoral (antibody-mediated) immune response in a verte-brate animal, large quantities of antibody specific for the foreign antigen appear in the serum and a state of long-lasting immunity to this antigen is developed. The clonal selection hypothesis, formulated independently over thirty years ago by Talmage[30] and Burnet,[5] proposes that immunity results from the selective ac-tivation of individual B lymphocytes. Prior to antigen encounter each B cell is committed to the expression of a single antibody structure which it expresses as a cell surface receptor. Activation results from the direct binding of antigen to this receptor. Amplification of antigen-specific antibody results from the clonal prolifer-ation of B cells that can bind the antigen. Expression of serum antibodies and the generation of long-lasting immunity then result due to differentiation of members of the clone to antibody-secreting plasma cells and to long-lived "memory" cells, respectively. Thus, according to the original clonal selection hypothesis, the devel-opment of humoral immunity is explained by a quantitative increase in a subset of a large repertoire of antibodies that pre-exist antigen encounter.

Several years after the formulation of the clonal selection hypothesis, obser-vations necessitated the alteration of the hypothesis. A number of investigators documented that not only the quantity but also the quality of antigen-specific antibody changed during the course of an immune response.[8,9,11,17,26,29,34] It was observed that the affinity of serum antibodies for the eliciting antigen steadily increased with time after immunization. To accommodate this phenomenon of "affinity maturation," amendments to the original hypothesis were made by a number of investigators. Most notable among these was the modification of Siskind and Benacerraf,[29] who proposed that the antigen selective mechanism was influenced by the affinity of the antigen-antibody interaction, and that antigen selection took place throughout the course of the response. Given these assumptions, and the reasonable conclusion that antigen concentrations steadily decline with time after immunization, the immune response was viewed as an ongoing and progressively stringent "competition" for available antigen among the B-cell clones stimulated at the onset of the response.

With the advent of recombinant DNA, nucleic acid sequencing, and hybridoma technologies, a quantum leap was taken in our understanding of the molecular and cellular basis of humoral immunity. Initial work in this area demonstrated that the potential of the immune system for generating diverse antibody structures was enormous and that a variety of distinct genetic mechanisms were responsible for this potential.[14,31] With these insights in hand, a number of laboratories embarked on experiments designed to evaluate how antibody structural diversity was generated and utilized during different stages of the immune response.[1,7,13,19,33] In such experiments, B cells were immortalized in the form of hybridoma cell lines before, at various times after primary immunization, and after secondary immunization of mice with a model antigen. Such hybridomas provided monoclonal sources of antibodies, representative of those expressed at various stages of the response, for structural analysis. These studies revealed that the clonal composition of the responding B-cell population changes with time, in accord with the predictions of Siskind and Benacerraf. More surprisingly, they also revealed that the V regions (the antigen-binding domains of antibodies) expressed by hybridomas derived several weeks after primary immunization, or several days after secondary immunization were extensively mutated, while the V regions expressed by hybridomas isolated before, or early after primary immunization were not. Thus, the process of V-region gene somatic hypermutation, which had been initially documented by Weigert and Cohn,[32] and subsequently by several groups of investigators,[3,6,28] apparently occurred exclusively during the immune response.

These new data demanded further modification of the clonal selection hypothesis. The most straightforward amendment that accommodated these data was a neo-Darwinian one—somatic mutation and antigen affinity-based clonal selection took place as stepwise functions of B-cell division.[20] Therefore, the immune response could be described as "evolution in a microcosm" where the individual members of a B-cell clone express different mutant forms of their V regions, and "compete" for the antigen required for further proliferation and mutation. The revelation that V-region mutation occurs at a high rate during the course of clonal expansion gave way to a scenario in which V regions follow certain mutational pathways to achieve high affinity for antigen, and that the boundaries of these pathways are defined by "mutational death" (loss of affinity for antigen).

Given the above introduction, it is apparent that the maturation of the immune response can be simulated as a large number of V regions "walking" on antigen affinity landscapes, where mutation allows a V region to move around on the landscape, and the topology of a landscape is defined by the manner in which the V region's affinity for antigen is influenced by this mutation. In two extreme scenarios, previously discussed by Kauffman,[15] if all mutations influence the affinity of a V region independently, its affinity landscape consists of a single peak. In contrast, if different mutations act synergistically or antagonistically in altering affinity, then a landscape is "rugged." Clearly, the adequate evaluation of such landscape models requires that far more be learned about the manner in which V domains respond to mutational alteration. Nevertheless, they provide a conceptual framework within which to discuss the "maturation" of the humoral immune response, and to help in the design of future experiments. Not surprisingly, landscape modeling studies have previously assumed that the maturation of the immune response takes place in a neo-Darwinian fashion,[16,24] that is, the mutation process is rather random (in terms of where mutations occur in a V region), and both mutation and antigen selection occur as stepwise functions of B-cell division. The data and discussion that follow call these assumptions into question.

CLONALLY RELATED SETS OF HYBRIDOMAS AND THE CONSTRUCTION OF SOMATIC PHYLOGENETIC TREES

Detailed evaluation of the mutational and antigen-selective events that take place during the maturation of the immune response requires that *in vivo* precursor-product relationships be defined between V regions with certain affinities for antigen, and their mutant derivatives with either higher, unchanged, or lower affinity for antigen. In this regard, the initial isolation of groups of hybridomas derived from single clones of B cells by Weigert's[25] and Rudikoff's[27] groups seemed promising. Sequence analysis of the V regions expressed by members of these clones revealed that mutations were shared to varying degrees among the members. Utilizing the doctrine of mutational parsimony, i.e., that the degree to which a mutation is shared among the offspring of a single founder is inversely related to the number of the generation in which this mutation took place, Weigert and Rudikoff constructed clonal phylogenetic trees from their hybridoma data. These constructions provided support for the idea that mutation occurs as a stepwise function of B-cell division, a prediction of the neo-Darwinian model for the maturation of the immune response.

In an effort to further test the neo-Darwinian model, and the conclusions drawn by Weigert, Rudikoff and others concerning the significance of V-region mutational patterns revealed by clonally related sets of hybridomas, we isolated a group of 20 such hybridomas from a single A/J mouse at an intermediate stage of the anti-iodo-arsonate (IArs) response. The data gathered in support of the conclusion that these hybridomas were derived from one clone of B cells have been reported elsewhere[22] and will not be discussed here. Sequencing of the V genes expressed by these hybridomas revealed that they contained somatic mutations, that mutations were shared to varying extents among the members of the clone, and that no single V domain shared all mutations with any other. However, attempts to construct phylogenetic trees from these data, following the doctrine of mutational parsimony, met with immediate difficulty—the most shared mutation among all of the hybridomas creates, in and of itself, an amber translation termination codon (indicated with an asterisk over codon 59 in Figure 1).

According to the neo-Darwinian model, if the criterion of maximum mutational parsimony is to be satisfied in this case, it must be assumed that a B cell that does not express an antigen receptor (due to the "amber" mutation) can nevertheless give rise to a large number of mutationally distinct progeny. Given this assumption, and the observation that none of the members of the clone actually fail to express a V region since in all cases one of three different additional mutations is present in the codon in which the "amber" mutation has occurred, the phylogenetic tree shown in Figure 2 can be constructed. In this tree the "amber" mutation is indicated by a double asterisk and the three distinct mutations that result in "reversion" of this mutation are indicated by exclamation points.

The conclusion that a B cell that lacks a functional antibody can continue to undergo extensive proliferation and differentiation is in conflict with the Siskind and Benacerraf modification of the clonal selection hypothesis stating that antigenic selection is an ongoing process during the immune response. Within the confines of their hypothesis, it must be assumed that the "amber" mutation could not occur alone, or the cell that expressed this mutation would cease to proliferate. Given this assumption, the phylogenetic tree shown in Figure 3 can be constructed. This tree assumes that second site mutations must have occurred in codon 59 either before, or simultaneously, with the "amber" mutation, such that a termination condon was never present. This assumption results in a tree in which the "amber" mutation takes place seven independent times.

Neither of the trees shown in Figures 2 and 3 are likely to accurately reflect the temporal order of the mutational events that occurred during the expansion of the clone *in vivo*. In both cases the criterion of mutational parsimony has been poorly satisfied—at least twenty of the 88 distinct V-region mutations observed must be assumed to have occurred independently multiple times during the expansion of the clone. Two possible explanations for this high frequency of "parallel" mutations that are within the confines of the neo-Darwinian model are apparent; either these "parallel" mutations represent mutational "hotspots," or these mutations confer significant increases in affinity for antigen—resulting in the recurrent selection of B cells that express these mutations.

```
        50                                    *    60
        TAT ATT AAT CCT GGA AAT GGT TAT ACT AAG TAC AAT GAG AAG TTC AAG
   1    --- --- --- --- --  -T- CT- --- --- T-C --- --- --- -G- --- ---
   2    --- --- --- --- -   --- --- --- --- --- --- --- --- -- --- ---
   3    --- --- --- --- -   --- --- --- --- TG- --- --- --- --A --- ---
   4    - -  -- --- --- --- -T- --- --- --- --- --- --- --- --- ---
   5    --- --- -- --- -    --- - - --- --- T-C --- -- --- -- -- ---
   6    --- --- --- --- -   -T- --- --- --- C-T --- --- - -  - --- ---
   7    --- --- --- --- -   --- --- --- --- T-C -C- --- --- -G  -- ---
   8    --- -   --- --- --- --- --- --- --- T-C --- --- --- --- ---
   9    --- --- --- --- --  --  --- -- --- TG- --- --- --- -C --- ---
  13    --- --- T-- --- --- --- -A- --- --- T-C --- --- --C --C --- -G-
  14    --- -A- --- --- --  --- --- --- --- T-T --- --- --- --- ---
  15    --- -- --- --- --   -- --- --- --- T-C --T --- --- --- ---
  16    --- --- --- --- --- G-- - - --- -TA T-C --- --- --- --- ---
  18    --- --- --- --- --- --- --- --- --- TG- --- --- --- --- ---
  19    --- --- --- --- --  --- --- --- -T- --- --- --- --- --- --T ---
  21    --- --- --- --- --  --- -- --- -T- C-C --- --- --- --- ---
  22    --- --- --- --- --- --- --- --- --- T-C --- --- -G- -- -- ---
  23    --- --- --- --- --  --- -A- --- -T- -T- --T --- --- --A --- ---
  25    --- --- --- --- --  --- -- --- CT- -T- --- --- --- -- --- ---
  26    --- --- --- -- --  G - -A- --- --- C-- --- --- --- -- --- ---
```

FIGURE 1 Nucleotide sequences in the CDR2 region of the V_H genes expressed by the clonally related hybridomas are presented. Each sequence is numbered, representing the hybridoma from which it was derived. Sequences are shown in comparison to the germline (unmutated) sequence of the encoding V_H gene segment. Dashes represent sequence identity. Gaps represent nucleotides that could not be unambiguously determined due to sequencing artifacts. Nucleotide differences are shown explicitly. The nucleotide that has been mutated to generate the "amber" codon is indicated with an asterisk above the reference sequence. Codon numbers are also indicated above the reference sequence (numbered sequentially from the mature amino terminus). The data are modified from those presented in Manser.[22]

The data presented in Figure 4 indicate that the mutational "hotspot" hypothesis is not a sufficient explanation. The affinities of 19/20 of the antibodies expressed by individual members of the clone are equal to or greater than the affinity of an antibody (indicated by the arrow in Figure 4) that represents the unmutated precursor to all of these antibodies. This demonstrates that extensive affinity maturation has taken place during the growth and differentiation of the clone and that many of the "parallel" mutations must confer affinity increases (this is known to be true for the combination of mutations that convert the position 59 AAG codon to TAC resulting in a lysine to tyrosine amino acid substitution.[10])

The analysis presented in Table 1 indicates that the neo-Darwinian "selection" explanation is also not sufficient to account for the data. The ratio of observed mutations that cause amino acid replacements (R) to those that do not (S—silent mutations) in subregions (CDR—complementarity-determining region, the portions of a V region that form the antigen binding site; and *FW*—framework, the portions of V-regions that confer their "superstructure") of the heavy chain V region genes expressed by members of the clone are not significantly different from the ratios expected if mutation had occurred randomly. Nevertheless, the frequency of mutation in *FW* subregions is ten-fold lower than in CDRs. If neo-Darwinian antigenic selection were responsible for this skewed distribution of mutations, the majority of mutations in *FW* should be *S* mutations, but they are not. Further, long periods of selection in the absence of mutation would presumably be necessary to produce the dearth of mutation observed in *FW* regions. However, the fact that no two of the V domains contain identical somatic mutations suggests that mutation was ongoing at the time the clone was immortalized as hybridomas.

The most significant conclusion that can be drawn from the data is that the efficiency with which this single clone of B cells sustained and fixed mutations that confer increased affinity for antigen was enormous. The high efficiency of the mutation-selection process that acted upon this clone is made even more striking by the following considerations: 1) the frequency of hybridoma formation from activated B cells is low—perhaps 0.1%[4]—so each of the independent mutation and selection events deduced from the hybridoma data must be multiplied by a "correct factor" of perhaps 1000; 2) the hybridomas were made at an intermediate stage of the immune response (day 16 after immunization) and mutation was probably not active for this entire period[21]; and 3) the increases in affinity observed among the antibodies are similar to those observed among antibodies isolated several months after immunization.[22,23]

According to a neo-Darwinian clonal selection model such efficiency can only be accounted for if it is assumed that: 1) mutations occur in CDRs more frequently than in FWs; 2) CDR mutational "hotspots" have a high probability of conferring increased affinity for antigen; and 3) mutation and selection are active over long periods of clonal expansion. While mutational "hotspots" have been documented in V-region genes,[2,12,18] it is difficult to imagine that a major fraction of the mutations that take place at such positions confer increased affinity for a specific antigen. If only a minor fraction of mutations result in an increase in affinity, then an unreasonably large number of cell divisions influenced by antigenic selection would be required to produce the nearly uniform dominance of high-affinity antibodies observed to be expressed by members of the clone.

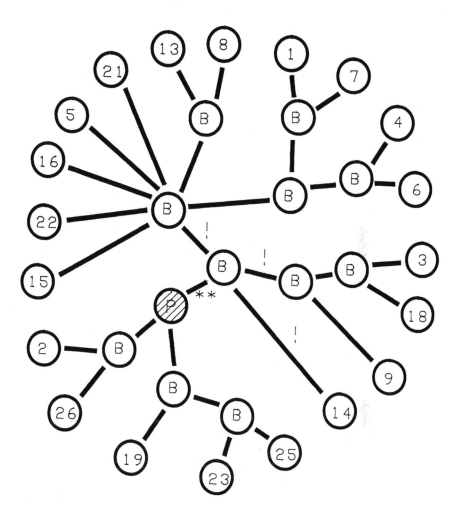

FIGURE 2 A phylogenetic tree generated from the sequences of the V genes expressed by the clonally related hybridomas assuming that the "amber" mutation occurred first and was followed by further proliferation and mutation. Circles containing numbers represent each of the hybridomas, circles containing B represent hypothetical mutational "branch point" cells, and the cross-hatched circle contain a P represents the hypothetical unmutated precursor to all of these cells. The "amber" mutation is indicated by a double asterisk and "reversion" mutations are indicated by exclamation points. The length of a branch is not proportional to the number of mutations that define that branch.

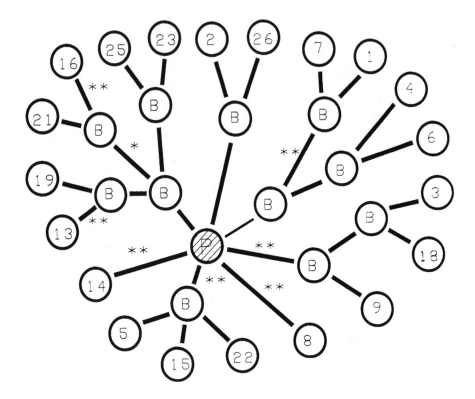

FIGURE 3 A phylogenetic tree generated from the sequences of the V genes expressed by the clonally related hybridomas assuming that the "amber" mutation could not be present alone. The tree is presented as described in the legend to Figure 2. Mutations that occurred in the same amino acid codon as the "amber" mutation, but took place before the "amber" mutation, thus preventing the formation of a termination codon, are indicated by a single asterisk.

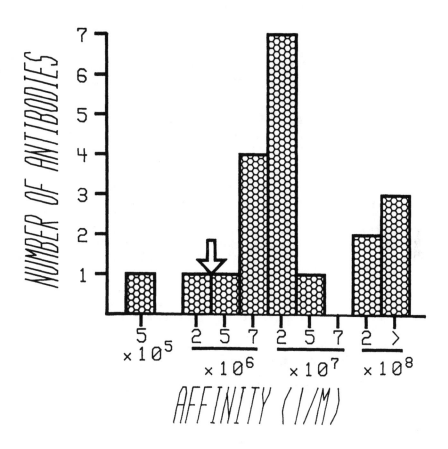

FIGURE 4 The intrinsic affinities of the antibodies expressed by each of the clonally related hybridomas for IArs, shown in histogram form. The data are a modification of those presented in Manser.[22] The affinities (presented as $K_{a'}$, in liters/Mol) are presented in groups that differ by less than three-fold (e.g., $1 - 3 \times 10^6$). The arrow indicates the affinity of a V region that exactly represents the hypothetical unmutated precursor to all of the V regions expressed by the clonally related hybridomas. The highest affinity group of antibodies is represented by a "$> \times 10^8$" since their high affinities could not be accurately measured due to small errors in functional antibody concentration.

TABLE 1 Distribution and type of mutations in the framework and complementarity determining regions of the V_H genes expressed by clonally related hybridoma.

	Exptected R/S Ratio	Observed R/S Ratio	Total Mutations	Frequency of Mutation
FWs	3.3	6.0	14	0.25
CDRs	4.5	6.25, 4.7	84, 55	3.3, 2.3

[1] A compilation of the mutations observed in the V_H genes expressed by the group of clonally related hybridomas and the ratio of mutations in framework (FW) and complementarity determining regions (CDR) that cause amino acid replacements in the encoded protein (R) to those that do not (S). In addition, the R/S ratios were calculated in the subregions assuming that mutation were to occur randomly (Expected R/S Ratio). Where two values are present under a given heading the first value is the number of mutations observed assuming that all mutations were independent events, and the second value in the number of mutations assuming that mutations that are shared among different V_H genes only occurred once. The frequency of mutation was calculated on a per nucleotide basis. The data are taken from Manser.[22]

A "NEO-INSTRUCTIONIST" MODEL FOR THE MATURATION OF THE IMMUNE RESPONSE

In toto, the data concerning the distribution and type of mutations in the V regions expressed by members of this set of hybridomas, and the affinities for antigen of these V regions, cannot be easily explained by even an extensively modified form of the neo-Darwinian clonal selection model. The data therefore call the validity of this model into question. While further experiments will be required to resolve this issue, it is perhaps constructive to begin to develop alternative models, and to consider experiments that could be done to test them. Following the doctrine of Ocham's razor is usually the method of choice when formulating hypotheses regarding complex phenomenon such as the maturation of the immune response. Figure 5 illustrates a model developed with these considerations in mind.

In this model V-gene mutation is assumed to occur independently of chromosomal DNA replication (and, thus, independently of cell division), and on only one strand (V2) of a local DNA replication bubble centered around the V-region gene. This results in the capacity of a single B cell to sequentially generate many distinct mutant forms (M, M+1, etc.) of its expressed V region during the course of a single cell cycle via multiple rounds of mutation and reversion (reversion being mediated by the DNA strand not subject to mutation). Antigenic selection is assumed to exert its action by inducing chromosomal DNA replication when the B cell generates a mutant V region with a certain affinity for antigen—the lower the amount of available antigen, the higher the requisite affinity. Since the DNA in and around the V gene has already been replicated, chromosomal DNA replication and subsequent cell division will result in the segregation of the two copies of the V gene (V1 and V2) to different daughter cells, one of which now contains only the mutant form (V2). The cycle may then be repeated in this cell, utilizing the new form of the V gene as the template for mutation. As antigen concentration progressively decreases during the response there is a commensurate increase in the affinity for antigen necessary to induce chromosomal DNA replication and cell division, resulting in the affinity maturation of antibodies expressed by the clone.

This new model is a selective model, however, the units of selection are single antibodies, not single cells. Since the antigen "instructs" individual B cells regarding which of the many mutant forms of their V regions can bind sufficient amounts of antigen and should therefore be passed on to daughter cells, but not regarding the specific types of mutation that should take place in a V region to generate high affinity, the model might be labelled "neo-instructionist." The appeal of this model in light of the data discussed here is that it predicts not only that the efficiency of affinity maturation on a per-cell basis will be high, but that deleterious mutations (e.g., the "amber" mutation) can be temporary and need not automatically lead to cell death.

There are several implications of this model for potential attempts to simulate the maturation of the immune response via landscape theory. As a V region "walks" across an affinity landscape it will take "downward" and "sideways" steps. According to the neo-Darwinian model, the time necessary to take a step is determined by the rate of cell division. When a "downward" or "sideways" step is taken, it is not easily reversed, and for every such step time to clonally expand is lost relative to other clones. The "neo-instructionist" model predicts that many steps per cell division may be taken and that a cell can ultimately do no worse then the V region with which it began the mutation-selection process.

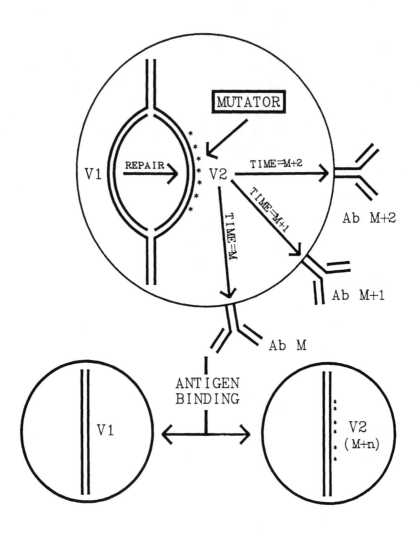

FIGURE 5   Representation of a "neo-instructionist" model for the somatic mutation—affinity-based antigen selection process. A single B lymphocyte (large circle) locally replicates the DNA centered around its productively rearranged V gene (V_H and/or V_L) giving rise to two copies (V1 and V2) of the gene. The somatic hypermutation mechanism (mutator) acts only on the V2 gene copy. This requires that the mutator can distinguish between the two copies of the V genes. This might result from differences in methylation of the two strands before replication. Alternatively, the mutator might affect only leading or lagging strand DNA synthesis. If this were true, and replication was initiated to one side of the V gene, only one of the replicating strands would suffer mutations. (continued)

FIGURE 5 (continued) In the first round of mutation a mutant form of V2 (a mutation is symbolized by an asterisk) is created and expressed to generate mutant antibody M. If both V1 and V2 were expressed the model could still explain the data, but if only V2 is expressed affinity maturation would operate more efficiently. Mutation is assumed to occur with roughly equal probability throughout the length of the V gene (but not necessarily with equal probability at each base pair in the gene). The model does not predict the nature of mutation itself, single or multiple nucleotide replacements might occur during each round of mutation and expression. In the next round the V2 gene may be repaired using germline information maintained in V1 and/or is then mutated again at time $M + 1$ giving rise to mutant antibody $M + 1$. The repair cycle may involve the entire V gene or only subregions of this gene. Such directional repair would require that the two strands be can be distinguished by repair mechanisms (see above). If any of the mutant antibodies $(M, M + 1, M + 2, \ldots, M + n)$ binds a stimulatory amount of antigen (see text) the B cell replicates the rest of its DNA and divides. The V1 and V2 copies of the original V gene are therefore segregated into two different daughter cells which can then locally replicate their V genes and begin further rounds of mutation. It should be noted that the cell inheriting the V2 gene begins this process with only a mutant gene as a template, that is all of the mutations in V2 are now fixed (cannot be reverted by germline information), while the cell that inherits V1 is identical to the cell that began the process (i.e., maintains a copy of the germline V gene).

CONCLUSIONS

More data are required before the "neo-Darwinian" and "neo-instructionist" models discussed above can be constructively compared and evaluated. In particular, several parameters regarding the maturation of the antibody response that have previously only been estimated must be accurately measured. Among these are: 1) the rate of B-cell division during an immune response *in vivo* and the influence of bound antigen concentration on this rate; 2) the size of expanded B-cell clones at various times during the response; and 3) the relative time frames in which somatic mutation and antigenic selection take place. Using the data currently available, theoretical exercises such as landscape modeling may help to delineate the "reasonable extremes" of such parameters, thereby allowing future experiments aimed at testing the better-designed models.

ACKNOWLEDGMENTS

This work was supported by grants from the NIH (AI23739) and the Pew Memorial Trust (Pew Scholars in the Biomedical Sciences). I would like to thank all members of the Manser laboratory for critical and constructive discussion.

REFERENCES

1. Berek, C., G. M. Griffiths, and C. Milstein. "Molecular Events During Maturation of the Immune Response to Oxazolone." *Nature* **316** (1985):412–418.
2. Berek, C., and C. Milstein. "Mutation Drift and Repertoire Shift of the Immune Response." *Immunol. Rev.* **96** (1987):23–42.
3. Bothwell, A. L. M., M. Paskind, M. Reth, T. Imanishi-Kari, K. Rajewsky, and D. Baltimore. "Heavy Chain Variable Region Contribution to the NP^b Family of Antibodies: Somatic Mutation Evident in a $\gamma 2a$ Variable Region." *Cell* **24** (1981):625–637.
4. Bruggeman, M., H.-J. Muller, C. Burger, and K. Rajewsky. "Idiotype Selection of an Antibody Mutant with Changed Hapten Binding Specificity, Resulting from a Point Mutation in Position 50 of The Heavy Chain." *EMBO J.* **5** (1986):1561–1570.
5. Burnet, F. M. "The Clonal Selection Theory of Acquired Immunity." London: Cambridge University Press, 1959.
6. Crews, S., J. Griffin, H. Huang, K. Calame, and L. Hood. "A Single V_H Gene Segment Encodes the Immune Response to Phophorylcholine: Somatic Mutation is Correlated with the Class of the Antibody." **25** (1981):59–66.
7. Cumano, A., and K. Rajewsky. "Clonal Recruitment and Aomatic Mutation in the Generation of Immunological Memory to the Hapten NP." *EMBO J.* **5** (1986):2459–2470.
8. Eisen, H. N., and G. W. Siskind. "Variations in Affinities of Antibodies During the Immune Response." *Biochem.* **3** (1964):996–1008.
9. Eisen, H. N. "The Immune Response to a Simple Antigenic Determinant." *Harvey Lectures*, vol. 60. New York: Academic Press, 1966, 1–33.
10. Fleming, M., S. Fish, J. Sharon, and T. Manser. Submitted.
11. Goidl, E. A., W. E. Paul, G. W. Siskind, and B. Benacerraf. "The Effect of Antigen Dose and Time After Immunization on the Amount and Affinity of Anti-Hapten Antibody." *J. Immunol.* **100** (1968):371–375.
12. Golding, G. B., P. J. Gearheart, and B. W. Glickman. "Patterns of Somatic Mutations in Immunoglobulin Variable Genes." *Genetics* **115** (1987):169–180.
13. Griffiths, G. M., C. Berek, M. Kaartinen, and C. Milstein. "Somatic Mutation and Maturation of the Immune Response to 2-Phenyl Oxazolone." *Nature* **312** (1984):271–275.

14. Honjo, T. "Immunoglobulin Genes." *Ann. Rev. Immunol.* 1 (1983):499–530.
15. Kauffman, S. A., and E. D. Weinberger. "Application of NK Model to Maturation of Immune Response." *J. Theoret. Biol.* 141 (1989):211–245.
16. Kauffman, S. A., E. D. Weinberger, and A. S. Perelson. "Maturation of the Immune Response Via Adaptive Walks on Affinity Landscapes." *Theoretical Immunology, Part One*, SFI Studies in the Sciences of Complexity, vol. II, edited by A. S. Perelson. Reading, MA: Addison-Wesley, 1988, 349–382.
17. Klinman, N. R., J. H. Rockey, G. Frauenberger, and F. Karush. "Equine Anti-Hapten Antibody III. The Comparative Properties of γG- and γA-Antibodies." *J. Immunol.* 96 (1966):587–595.
18. Levy. S., E. Mendel, E. Kon, E. Shinichiro, Z. Avnur, and R. Levy. "Mutational Hot Spots in Ig V Region Genes of Human Follicular Lymphomas." *J. Exp. Med.* 168 (1988):475–490.
19. Malipiero, U. V., N. S. Levy, and P. J. Gearhart. "Somatic Mutation in Anti-Phosphorylcholine Antibodies." *Immunol. Rev.* 96 (1987):59–75.
20. Manser, T., L. J. Wysocki, T. Gridley, R. I. Near, and M. L. Gefter. "The Molecular Evolution of the Immune Response." *Immunol. Today* 6 (1985):94–101.
21. Manser, T., L. J. Wysocki, M. N. Margolies,and M. L. Gefter. "Evolution of Antibody Variable Regions Structure During the Immune Response." *Immunol. Rev.* 96 (1987):141–162.
22. Manser, T. "Evolution of Antibody Structure During the Immune Response: The Differentiative Potential of a Single B Lymphocyte." *J. Exp. Med.* 1989.
23. Manser, T. Unpublished observation
24. Macken, C. A., and Perelson, A. S. "Protein Evolution on Rugged Landscapes." *Proc. Natl. Acad. Sci. USA* 86 (1989):6191–6195.
25. McKean, D., K. Huppi, M. Bell, L. Staudt, W. Gerhard, and M. Weigert. "Generation of Antibody Diversity in the Immune Response of BALB/c Mice to Influenza Virus Hemagglutinin." *Proc. Natl. Acad. Sci. USA* 81 (1984):3180–3185.
26. Parker, C. W., S. M. Gott, and M. C. Johnson. "The Antibody Response to a 2,4-Dinitrophenyl Peptide." *Biochem.* 5 (1966):2314–2326.
27. Rudikoff, S., M. Pawlita, J. Pumphrey, and M. Heller. "Somatic Diversification of Immunoglobulins." *Proc. Natl. Acad. Sci. USA* 81 (1984):2162–2167.
28. Selsing, E., and U. Storb. "Somatic Mutation of Immunoglobulin Light-Chain Variable Region Genes." *Cell* 25 (1981):47–58.
29. Siskind, G. W., and B. Benacerraf. "Cell Selection by Antigen in the Immune Response." *Adv. Immunol.* 10 (1968):1–50.
30. Talmage, D. W. "Allergy and Immunology." *Ann. Rev. Med.* 8 (1957):239–250.
31. Tonegawa, S. "Somatic Generation of Antibody Diversity." *Nature* 307 (1983):575–580.
32. Weigert, M. G., I. M. Cesari, S. J. Yonkovich, and M. Cohn. "Variability in the Lambda Ligh Chain Sequences of Mouse Antibody." *Nature* 228 (1970):1045–1047.

33. Wysocki, L., T. Manser, and M. L. Gefter. "Somatic Evolution of Variable Region Structures During an Immune Response." *Proc. Natl. Acad. Sci. USA* **83** (1986):1847–1852.
34. Zimmering, P. E., S. Liberman, and B. F. Erlanger. "Binding of Steroids to Steroid-Specific Antibodies" *Biochem.* **6** (1967):154–164.

Stuart A. Kauffman†‡ and Edward D. Weinberger‡
‡University of Pennsylvania, Philadelphia, PA 19104-6059 and †Santa Fe Institute 1120
Canyon Road, Santa Fe, NM 87501

The NK Model of Rugged Fitness Landscapes and Its Application to Maturation of the Immune Response

This paper originally appeared in the *Journal of Theoretical Biology*, volume 141, pages 211–245 (1989). It is reprinted by permission.

Adaptive evolution is, to a large extent, a complex combinatorial optimization process. Such processes can be characterized as "uphill walks on rugged fitness landscapes." Concrete examples of fitness landscapes include the distribution of any specific functional property such as the capacity to catalyze a specific reaction, or bind a specific ligand, in "protein space." In particular, the property might be the affinity of all possible antibody molecules for a specific antigenic determinant. The affinity landscape presumably plays a critical role in maturation of the immune response. In this process, hypermutation and clonal selection act to select antibody V-region mutant variants with successively higher affinity for the immunizing antigen. The actual statistical structure of affinity landscapes, although knowable, is currently unknown. Here, we analyze a class of mathematical models we call NK models. We show that these models capture significant features of the maturation of the immune response, which is currently thought to share features with general protein evolution. The NK models have the important property that, as the parameter K increases, the "ruggedness" of the NK landscape varies from a single-peaked "Fujiyama" landscape to a multi-peaked "badlands" landscape. Walks to local optima on such landscapes become shorter as K increases. This fact allows us to choose a value of K that corresponds to the observed number of mutation steps on an adaptive walk to a local optimum. Tuning the model requires that K be about 40. Given this value of K, the model then *predicts* several features of "antibody space" that are in qualitative agreement with experiment: (1) The fraction of fitter variants of an initial "roughed in" germline antibody amplified by clonal selection is about 1–2%. (2) Mutations at some sites of the mature antibody hardly affect antibody function at all, but mutations at other sites dramatically decrease function. (3) The same "roughed in" antibody sequence can "walk" to many mature antibody sequences. (4) Many adaptive walks can end on the same local optimum. (5) Comparison of different mature sequences derived from the same initial V region shows evolutionary hot spots and parallel mutations. All these predictions are open to detailed testing by obtaining monoclonal antibodies early in the immune response and carrying out *in vitro* mutagenesis and adaptive hill climbing with respect to affinity for the immunizing antigen.

INTRODUCTION

The evocative imagery created by Wright's notion of an adaptive landscape[61] is one of the most powerful concepts in evolutionary theory. The simplest version of his idea pictures a space of genotypes, each "next to" those other genotypes which differ by a single mutation, and each assigned a fitness. The distribution of the fitness values over the space of genotypes constitutes the fitness landscape. Maynard-Smith[39] borrowed Wright's image in defining "protein space" and adaptive walks in that space. The space consists of all 20^N proteins, length N, arranged such that each protein is a vertex next to all of the $19N$ 1-mutant variants obtained by replacing one amino acid at one position with one of the 19 remaining possible coded amino acids. Each protein in the space is assigned some "fitness" with respect to a specific property; such an adaptive walk on a fixed landscape must climb to a locally optimal protein, better than all $19N$ 1-mutant variants. These ideas can be generalized to walks proceeding via 2-mutant or more distant neighbors, or those allowed to pass via less-fit neighbors as well. More recently, other authors have taken up the idea of protein space, or more generally of sequence space.[15,21,33,34,35,36,43,49]

It is clear that the character of such adaptive walks depends upon the actual structure of the fitness landscape, whether it is smooth with few adaptive peaks, or highly mountainous and multi-peaked. In addition the adaptive process depends upon the actual mechanisms of adaptive "flow" by the population: for an asexual population these include the mutation rate and population size. Gillespie[26,27] has shown that in the limit where the mutation rate is low and the relative fitness differences between less-fit and more-fit mutant neighbors is sufficiently great, the general flow of a population over a landscape can be simplified to a process where the population as a whole remains fixed at a single "genotype," protein, or point in the space for long times, then moves as a unit to a fitter 1-mutant variant. We use this limiting case of an adaptive walk in this article, because it allows us to focus on the statistical structure of rugged adaptive landscapes. Nevertheless, the actual flow of a population of maturing B cells on the real affinity landscape is very likely to be a more complex process which "spreads out" along fitness ridges in the affinity landscape in ways depending upon the details of mutation rate, cell population sizes, and fitness differences. Once the statistical structure of affinity landscapes are understood, these further issues must also be addressed.

The actual structure of fitness landscapes in protein space for specific catalytic or ligand-binding functions is unknown, but increasingly open to direct investigation by current genetic engineering and site-directed mutagenesis studies. Our aim in this article is to discuss further a spin-glass-like model of random epistatic interactions, called the NK model, introduced and considered elsewhere.[33,34,36] N is the number of "sites" in the model genotype or protein, while K is the number of sites whose alternative states, "alleles" or amino acids, bear on the fitness contribution of each site. Thus K measures the richness of epistatic interactions among sites.

This model generates a family of increasingly rugged multipeaked landscapes as its main parameters are tuned.[33,34,36] Thus as K increases relative to N, landscapes pass from smooth and single peaked to jagged and multi-peaked. Of course, since protein space is a discrete "sequence space," the fitness values are only defined on this discrete space. The general interest in this family of landscapes lies in understanding the implications of the richness of epistasis on the expected structure of fitness landscapes. Thus the model can be interpreted as a haploid genetic model and used to study the effects of epistasis in population genetics. Our particular purpose in the present article is to show that the NK model predicts a number of features of a well-know example of rapid adaptive protein evolution: maturation of the immune response.

We wish to contrast our approach, which considers only the statistical structure of the landscape, with more familiar theoretical approaches that involve detailed simulation of actual protein molecules. We believe we are proposing a kind of "statistical mechanics" of the immune response in particular, and protein evolution in general. This analogy with physics seems apt, because the theory is motivated by a desire to understand the *ensemble* properties of evolution among proteins even at the risk of ignoring important details regarding individual proteins. There are two reasons for taking this position, one practical and the other theoretical. First, the practical reason: there is an extensive literature which discusses detailed mathematical models of proteins that include the main chain, electrostatic, van der Waals, and hydrogen-bonding forces, and, perhaps, forces due to interactions with a solvent.[31,32,52] Unfortunately, detailed analysis of the kind of adaptive walks in the space of antibodies which we propose is computationally intractable on current computers. From a theoretical perspective, we need to understand the actual statistical structure of fitness landscapes underlying protein evolution. If simple statistical models such as the NK model we discuss predict actual adaptive landscapes in protein evolution, then we may hope that the NK model or improved variants point to the underlying basis for the structure of protein adaptive landscapes. Such models may help teach us how proteins work and evolve.

The rest of this paper is laid out as follows: In the first section, we present a more detailed discussion of the idea of peptide spaces, which we choose to rename "affinity landscapes," to reflect our interest in the binding of antigens by antibodies. This leads us to note a number of natural features of mountainous fitness landscapes which are open to experimental and theoretical investigation. In the second section we discuss the NK class of mathematical models, and discuss enough of its properties to motivate the modeling steps employed subsequently. The third section sketches the biological facts regarding the maturation of the immune response. In that section, we suggest that this well-studied system, during which hypermutation and clonal selection amplify V-region mutant antibodies with successively higher affinity for the immunizing antigen, is a natural testbed for the application of the NK model to protein evolution. The fourth section summarizes the predictions of the model and its qualitative agreement with relevant experiments. We conclude with a discussion of the significance and limitations of our findings and some suggestions for avenues of future investigations.

THE STRUCTURE OF AFFINITY LANDSCAPES

In this section, we set forth the concept of an affinity landscape in more detail. The set of all 20^N proteins of length N can be represented as points in an abstract N-dimensional space in which proteins that differ at exactly one residue are "neighbors." Although it is difficult to draw a picture of such high-dimensional spaces, a sense of their structure can be captured by considering proteins with only two amino acids, e.g., alanine and glycine. In Figure 1(a) , all $2^4 = 16$ possible peptide sequences of length 4 for these two amino acids are shown, using the representation "1" for alanine and "0" for glycine. Each vertex in this figure corresponds to a specific sequence, e.g., "1101" or "0101." There are also four lines emanating from each vertex. These lines connect the vertex with the four other "neighbor" vertices that differ from the first by a single amino-acid substitution. It is thus clear from the figure that the neighbors of, e.g., "0110" are "1110," "0010," "0100," and "0111." Peptide sequences involving all 20 amino acids would be structurally similar, but more complicated to draw. Each dimension would have 20 vertices, and each vertex would be connected to the 19 others in the same hyperplane. Because there are N hyperplanes, there are $19N$ total connections for each vertex. The "protein space" construction allows us to specify exactly what we mean by neighboring sequences, the minimum number of changes to pass from one sequence to another, etc. We also remark in passing that the concept is very general, and can be used to represent entire organisms or other ensembles of related objects which are "one-mutant neighbors" of each other.

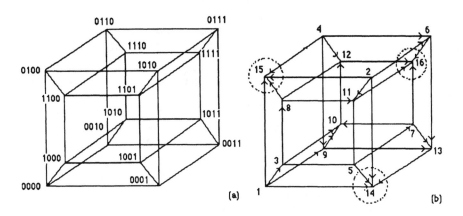

FIGURE 1   (a) A four-dimensional "Boolean hypercube" showing all 16 possible peptides, length four, comprised of only two amino acids, alanine =1, glycine =0. Each peptide is "next" to those which are accessible by mutating a single amino acid. (b) Each peptide has been assigned, at random, a rank-order fitness, one low, 16 high. Arrows connect adjacent peptides, and point to the peptide of higher fitness. Circles surround local optima in this small peptide space.

As noted above, we can assign a "fitness" to each protein by measuring its capacity to perform a specific function, such as catalyzing a given reaction, binding a given ligand, etc. In Figure 1(b), we have assigned each of the 16 peptides on the vertices with the hypothetical fitness values shown. This assignment gives a rank order to the peptides from the worst (1) to the best (16). As we will see, the properties of interest in the subsequent discussion depend only on these rank orderings. The adaptive walks we focus on might begin with any of the 16 peptides and will "step" to a 1-mutant neighbor peptide only if the second peptide is fitter (has higher rank order) than the first. In Figure 1(b) this is represented by arrows from each peptide directed to those 1-mutant neighbors with higher rank order. The sequence of arrows connecting a series of adjacent vertices in the Figure represent such a walk. The walk must terminate when it reaches a peptide which is fitter than all of its 1-mutant neighbors. Such a peptide is a *local optimum* of the space. In Figure 1(b) three of the 16 peptides are local optima.

To completely specify the character of the walk, it is necessary to choose a step selection mechanism. Two choices are natural. The first is to step to the neighbor with the highest fitness. Such walks are called *greedy walks*. The second involves selection the fitter neighbor *at random* from among all the fitter 1-mutant neighbors. Either of these two is an idealization of the actual flow of an adapting population on a rugged landscape under the drive of mutation and selection. The former roughly represents a case where more than a single mutant is encountered in a short time period and the fittest variant sweeps the population. The latter roughly represents the case considered by Gillespie: when the rate of finding a fitter variant is low compared to selection differences ensuring the rapid establishment of any fitter variant, once it is found and present insufficient numbers to obviate loss by chance drift. We will use Gillespie's limiting case and consider fitter 1-mutant variants to be chosen at random.

The simple landscape shown in Figure 1(b) has rank-order fitness values which were assigned at random to the 16 possible peptides in the space. In considering maturation of the immune response the natural measure of fitness is the *affinity* with which each possible antibody binds the immunizing epitope on the specific antigen with respect to which maturation is occurring. Since the diversity of antibodies in the mammalian immune response is thought to be greater than 10^8, while each antibody is a 1-mutant neighbor of thousands of other antibody molecules, affinity landscapes are far more complex.[5,29] Yet whatever the other cellular mechanisms may be which underlie hypermutation, clonal selection, and other parts of the maturation process, it seems obvious that the unknown mountainous structure of affinity landscapes in antibody space must be central to the maturation process.

The primary virtue of the landscape construct is that it raises a number of theoretically and experimentally accessible questions about the nature of uphill walks and the optima that they reach. All are evident in Figure 1:

1. How many local optima exist in a landscape?
2. What is the distribution of optima in the landscape? Are they near one another in special subregions of the space or randomly scattered?
3. What are the lengths of uphill walks to local optima?
4. As an optimum is approached, the fraction of fitter neighbors must dwindle to 0. How rapidly does the fraction of fitter neighbors dwindle?
5. Because the fraction of neighbors which are fitter dwindles to 0, there is some characteristic relation between the number of mutations "tried" and the number "accepted" on an adaptive walk. How are the two related?
6. How many alternative optima are accessible from a given starting point? Can a "low-fit" peptide typically climb to all possible local optima, or only a small fraction of those optima? Among the accessible alternative optima, how often will each be "hit" on independent adaptive walks from the same starting point?
7. How many of the possible peptides can climb to any specific optimum, including the global optimum? A small fraction? Almost all?
8. Since most adaptive walks end on local optima, what are the fitnesses of such optima and how do they compare with the global optimum in the space?
9. The 1-mutant variants of a local optimum must be less fit than the optimum. But do all of the variants lead to nearly the same loss of fitness or is there high variance indicating precipitous cliffs and gentle ridges in different directions in the high-dimensional space?

Previous work[35,37,60] has analyzed many of these questions with respect to the limiting case of *fully random landscapes* in which the fitness of 1-mutant neighbors are assigned at random from some fixed underlying distribution. In such an uncorrelated landscape, the fitness of one protein carries no information about the fitness of its 1-mutant neighbors. Presumably the real fitness landscapes underlying protein evolution are not uncorrelated, although they are correlated in as yet unknown ways. In order to begin to gain insight into the structure of *correlated* fitness landscapes, we previously introduced the NK family of rugged landscapes.[33,34,36] Our hope is that this family of correlated landscapes, characterized by a few major parameters, may make reasonable predictions about the actual structure of antibody affinity landscapes. We present the NK family next.

THE NK MODEL: A SPIN-GLASS-LIKE MODEL OF AN AFFINITY LANDSCAPE

The NK model is meant to apply to systems of many, N, parts, where the functional contribution of each part depends upon the "state," among A alternatives, of the part, and is epistatically affected by an average of K other parts. In the case of genotypes, the N parts are interpreted as genes, the A alternative states of a part of the alleles, and the K epistatic interactions as functional effects of the alleles at other loci upon the fitness contribution of an allele at a specific locus to overall fitness. In the case of a protein, the N parts are the amino acids in the primary sequence, the A states are the 20 possible amino acids, and K measures the average number of other sites in the primary chain whose amino acids bear on the functional contribution of the amino acid at a given site to overall function. In reality, K presumably varies from site to site. We treat it as a constant for the moment. In short, K is a parameter which measures how richly interconnected the parts of the system are. As we shall see, increasing K from 0 to $N-1$ increases the number of peaks and valleys, and thus the ruggedness of the corresponding fitness from single peaked and smooth to multi-peaked and fully uncorrelated. In turn, the ruggedness of the landscape alters the character of adaptive walks towards optima under biologically reasonable mutation selection models or any of a variety of optimization procedures. In addition to specifying the values of N, A, and K, it is also necessary to specify for each site the specific K among the N which affect it. For example, one might wish to assume reciprocity. If site I affects J, then J affects I. Alternatively, reciprocity might not be assumed. More generally, if the sites are located in a linear structure such as a chromosome or protein, the K sites bearing on any site might be its neighbors, chosen at random, or in some non-random spatial distribution.

The central idea used in the NK model is that the epistatic effects of the A^K different combinations of A alternative states of the K other sites on the functional contribution of Ath state of each part are so complex that their *statistical* features can be captured by assigning fitness consequences at *random* from a specified distribution. In this sense the NK model is a model of random epistatic interactions. Given N, A, K, the distribution of K among the N assigned to each site, and the underlying fitness distribution from which random fitness contributions assignments are made, the NK model is specified and in turn determines an ensemble of fitness landscapes. The model is similar to spin-glass models of disordered magnetic materials which have received extensive attention in solid state physics recently.[1,14,53] In fact, for the case where the K epistatic sites are a site's flanking neighbors, the model can rigorously be shown to be a type of short-range spin glass.[6] Conversely the case in which $K = N-1$ corresponds to the Derrida random-energy spin glass,[12] as becomes clear below.

Consider the simplest version of the NK model which assumes that each site has only $A =$ two states. This corresponds to an N-locus two-allele haploid genetic model, or, as in the previous section, to a restricted peptide space in which only two of the 20 biologically important amino acids are present (e.g., alanine $=0$ and glycine $=1$). Each amino acid makes a fitness contribution depending on whether it is 0 or 1, and whether the K other amino acids which impinge upon it are 0 or 1. Thus the fitness contribution of each of the N sites depends upon the state at $K + 1$ sites.

The NK model assigns a "fitness contribution," w_i in $(0, 1)$, to each amino acid, $i, 1 \le i \le N$, of the N-residue chain such that w_i depends on i and $K < N$ other bits. Since each amino acid can be 0 or 1, there are $2^{(K+1)}$ combinations of states of the $K + 1$ amino acids which determine the fitness contribution of each amino acid. The fitness contributions associated with each of these combinations

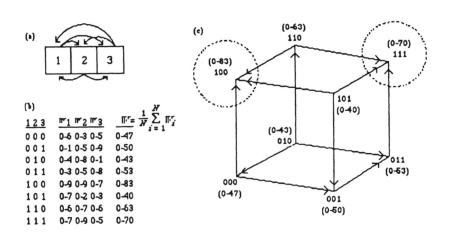

FIGURE 2   (a) Three adjacent sites in the NK model, each receives epistatic inputs from the other two; $N = 3, K = 2$. (b) The fitness contribution of each site, $w_i, i = 1, 2, 3$, as a function of the allele, 1 or 0, at that site and at the $K = 2$ other sites which bear upon it. The fitness, W, of each genotype, or tripeptide is the average of the fitness contributions of the three sites. (c) The fitness of each of the $2^3 = 8$ possible genotypes, or tripeptides, from 2(b), on the three-dimensional Boolean cube representing this small sequence space. Note that two local optima exist.

is assigned by selecting an independent random variable from the uniform distribution on $(0, 1)$. This constitutes the "fitness table" for the ith amino acid. There is a different, independently generated table for each of the N amino acids. Then, given any "string" of N amino acids, the total fitness of the string, W, is defined as the average of the fitness contributions of each part (i.e., each of the $w_i's$), each in the context of the K which impinge upon it:

$$W = \frac{1}{N} \sum_{i=1}^{N} w_i .$$

(1)

Figure 2(a) shows a simple example of the NK model for a hypothetical three-gene system, each with two alleles, where the fitness contribution of each allele at each locus depends on the alleles at the locus and the two other loci. Equivalently, this model, a tripeptide with three amino acids, each of which makes a contribution to overall function depending upon the amino acid at the site and the remaining two sites. As shown, the resulting fitness tables for the three sites (Figure 2(b)) yield a fitness for each of the $2^3 = 8$ possible tripeptides, and in turn induce a fitness landscape like that in Figure 1, with adaptive walks to local optima (Figure 2(c)).

The NK model affords a "tuneably rugged" fitness landscape, since tuning K alters how rugged the landscape is. This can be seen from the following: For $K = 0$, each site is independent of all other sites. Except for rare "ties" which we ignore, either the bit value 0 or the bit value 1 is "fitter" than the other; hence, a single specific sequence comprised of the fitter bit value in each position is the single, global optimum in the fitness landscape. This simple case corresponds to the familiar haploid, multilocus, two-allele-additive genetic model found in population genetics (see, e.g., Ewens[18]). The "correlation" of a fitness landscape measures how similar the fitness values of "1-mutant" variants in the space of systems are. Specifically, each N-bit sequence has N 1-mutant neighbors, obtained by mutating any bit to the opposite state. Since such a mutation can only alter fitness by $1/N$ or less in $K = 0$ landscapes, such landscapes are highly correlated. Furthermore, in such landscapes, the statistics of local hill-climbing walks are simply obtained. There is a single sequence which is the global optimum. Any other sequence is suboptimal, and is on a connected walk via 1-mutant fitter variants to the global optimum by mutating bits from less-fit to more-fit values. The length of the walk is just the number of bits by which the initial sequence differs from the global optimum. For a randomly chosen initial string, half the bits will be in their less-fit state, hence the expected walk length is just $N/2$. Thus the lengths of adaptive walks to the global optimum scale linearly with N. Further, at each adaptive step the number of fitter 1-mutant neighbors decreases by one, hence the directions uphill dwindle slowly as the global optimum is approached. These properties are in sharp contrast to the limit when $K = N - 1$.

The fully connected NK model yields a completely random fitness landscape. For $K = N - 1$, the fitness contribution of each site depends on all of the other sites in the sequence and therefore altering any site from one to the other value, 0-1, alters the fitness contribution of each site to a new random value. Thus the fitness of any 1-mutant neighboring sequence is completely random with respect to the initial sequence. The landscape is fully random. As was shown in Kauffman and Levin,[35] Weinberger,[60] and Macken and Perelson,[37] such random landscapes have very many local optima, on average, $2^N/(N+1)$. Walks to optima are short, scaling as ln N. The expected fraction of fitter neighbors falls by half at each adaptive step. The "time" or number of mutants "tried" to reach a local optimum scales as N, the number of 1-mutant neighbors of a sequence. Thus the ratio of accepted to tried mutations itself scales as $(\ln N/N)$.[33,34,37] Only a small fraction of local optima are accessible from any initial string, and only a small fraction of peptides can climb to any optimum, including the global optimum.[33,34,35] Finally, as N increases, and K remains equal to $N-1$, the *fitness of local optima fall* toward the mean of the space, $0 \cdot 5$, in a kind of *complexity catastrophe*. On the other hand, simulation results and some analytic work suggests that the fitness of the global optimum in the space appears not to fall as N increases.[33] In short, adaptive walks vary dramatically as the ruggedness of the landscape varies.

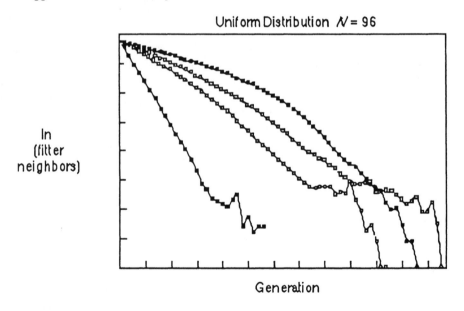

FIGURE 3 Logarithm of the mean number of fitter 1-mutant neighbors at each adaptive step, plotted against the adaptive step, or generation, for the NK model with $N = 96$, and fitness values chosen uniformly between 0 and 1. As K increases, the rate of fall in the fraction of fitter 1-mutant neighbors increases as well.

To gain insight into the behavior of the model for $0 < K < N - 1$ with two amino acids, we simulated 100 different instances of uphill walks for each of the N and K values given in Tables 1 and 2. Walks started from random initial states. These tables show the means and, in parentheses, the standard deviations of the maximal fitnesses attained, Tables 1(a) and 2(a), and mean walk lengths, Tables 1(b) and 2(b), for the cases in which the K sites are nearest neighbors, Tables 1(a) and (b), or randomly chosen, Table 2(a) and (b). The largest possible value of K is $N - 1$. For simplicity, in these tables we use $K = N$. Figure 3 shows that as K increases relative to N, the fraction of fitter neighbors at each adaptive step dwindles more rapidly. The reciprocal of the fraction of fitter neighbors is the expected waiting time to find a fitter variant, or equivalently the expected number of mutants "tried" to take the next adaptive step.

Note that as N increases for $K = N - 1$, the fitness of optima fall. The complexity "catastrophe" inexorably sets in. Indeed Tables 1(a) and 2(a) suggest that K need only increase linearly with N for this ultimate decline in the fitness of accessible optima to occur. Conversely Tables 1(a) and 2(a) indicate that if K remains small while N increases, local optima do not fall in fitness. This hints at a construction requirement: as the number of interacting pats in a complex system increase, the adaptive landscape will tend to retain highly fit and accessible local optima if the epistatic interactions remain low. Note also that as K increases relative to N, mean walk lengths to optima decrease. We use this fact below to apply the NK model to maturation of the immune response. Table 3 shows the number of local optima found for different values of N and K. An important general feature of these results is that the basic structure of the landscape is quite insensitive to whether the K sites affecting each site are its neighbors or assigned at random. This will recur in our application of the NK model to maturation of the immune response. We comment that for the limiting case of $K = N - 1, K = 15, N = 16$, the number of local optima encountered appears higher than the predicted $2^N/(N + 1)$. The reasons for this disparity are not clear.

NK landscapes have other quite striking properties, most but not all of which are quite insensitive to the detailed assumptions of the model, as discussed elsewhere.[33,34] For example, for $A = 2$ and $K = 2$, whether the epistatic sites are adjacent or random, the general configuration of the landscape is very non-random. The highest optima are both nearest one another and also have the largest drainage basins. The landscape has a *Massif Central*. As A and K increase these ordered features decay rapidly with respect to the nearness of high local optima to one another, more slowly for the tendency of the highest optima to drain the largest region of the space. Since these features and their implications are described elsewhere,[33] we do not comment upon them further here.

TABLE 1 (a) Mean fitness of local optima attained on walks on 100 different landscapes for each different value of N and K. Standard deviations shown in brackets. K sites bearing on each site were its $K/2$ flanking neighbors on each side. Circular sequences were assumed to avoid boundary effects. Two alternative "alleles" or amino acids are possible at each site.

			N		
K	8	16	24	48	96
0	0·65(0·08)	0·65(0·06)	0·66(0·04)	0·66(0·03)	0·66(0·02)
2	0·70(0·07)	0·70(0·04)	0·70(0·08)	0·70(0·02)	0·71(0·02)
4	0·70(0·06)	0·71(0·04)	0·70(0·04)	0·70(0·03)	0·70(0·02)
8	0·66(0·06)	0·68(0·04)	0·68(0·03)	0·69(0·02)	0·68(0·02)
16		0·65(0·04)	0·66(0·03)	0·66(0·02)	0·66(0·02)
24			0·63(0·03)	0·64(0·02)	0·64(0·01)
48				0·60(0·02)	0·61(0·01)
96					0·58(0·01)

TABLE 1 (b) Mean walks lengths to local optima attained on walks on 100 different landscapes for each different value of N and K. K adjacent.

			N		
K	8	16	24	48	96
0	1·5(1·2)	8·6(1·9)	12·6(2·2)	24·3(3·4)	48·8(5·6)
2	4·1(1·9)	8·1(3·2)	11·2(3·1)	22·5(4·6)	45·2(6·6)
4	3·2(1·8)	6·6(2·5)	9·4(2·9)	19·3(3·9)	37·3(6·1)
8	2·7(1·5)	4·7(2·3)	7·7(3·0)	15·3(4·3)	27·7(5·3)
16		3·3(1·7)	4·8(2·1)	9·6(3·0)	19·3(4·2)
24			3·5(1·4)	7·4(3·0)	5·0(3·9)
48				3·9(1·9)	8·9(3·0)
96					5·1(2·4)

TABLE 2 (a) As in Table 1(a), except the K sites bearing on each site were randomly chosen. Sequences were not assumed to be circular.

K	8	16	N 24	48	96
2	0·70(0·06)	0·71(0·04)	0·71(0·03)	0·71(0·03)	0·71(0·02)
4	0·68(0·05)	0·71(0·04)	0·71(0·04)	0·72(0·03)	0·72(0·02)
8	0·66(0·06)	0·69(0·04)	0·69(0·04)	0·70(0·02)	0·71(0·02)
16		0·65(0·04)	0·65(0·03)	0·67(0·03)	0·68(0·02)
24			0·63(0·03)	0·65(0·02)	0·66(0·02)
48				0·60(0·02)	0·62(0·02)
96					0·58(0·01)

TABLE 2 (b) As in Table 1(b), except the K sites bearing on each site were randomly chosen.

K	8	16	N 24	48	96
2	4·4(1·8)	8·1(2·8)	12·5(3·8)	26·5(5·1)	46·9(6·1)
4	3·6(1·8)	7·1(2·9)	10·9(3·3)	22·9(5·6)	44·5(7·9)
8	2·7(1·5)	5·3(2·5)	8·0(3·2)	17·0(4·3)	34·7(6·5)
16		3·3(1·7)	4·8(2·1)	10·1(3·4)	21·6(4·8)
24			3·5(1·4)	7·4(2·6)	16·0(4·3)
48				3·9(1·9)	9·3(2·6)
96					5·1(2·4)

TABLE 3 Number of optima
in landscapes for different
values of N and $K, K <$
N. Data are means of ten
landscapes, each explored
from 10,000 different initial
points.

K	N	
	8	16
2	5	26
4	15	184
7	34	—
8	—	1109
15	—	4370

CONCLUSIONS FOR THE N-LOCUS TWO-ALLELE OR TWO AMINO ACID CASE

The NK model is a very general approach to understanding complex epistatic interactions and their effects upon the structure of correlated but mountainous landscapes. The main conclusions to bring away are that increasing K relative to N increases the number of local optima, shortens the lengths of walks to optima, increases the rate at which fitter neighbors dwindle to 0 along adaptive walks, increases the ruggedness of the landscape, reduces the fraction of optima accessible from a given point, reduces the number of points which can climb to a given optimum, and leads to a complexity catastrophe in which accessible optima fall toward the mean of the space. All these features presumably reflect the fact that, as K increases, more conflicting constraints, or what in spin-glass models is called *frustration*,[1] sets in.

While the NK family of landscapes is of interest and appears to be one useful model of correlated fitness landscapes, there may be very many such families of landscapes. Ultimately we must be interested in the actual structure of the fitness landscapes underlying adaptive protein evolution. In the next section we describe maturation of the immune response, then ask whether the NK landscapes have approximately the right statistical features to fit the known features of that adaptive landscape.

MATURATION OF THE IMMUNE RESPONSE

The purpose of this section is to suggest that maturation of the immune response is the kind of adaptive process that we have described previously, so that the NK model has a natural interpretation in this context. We also show that our ideas about "adaptive walks on rugged landscapes" give rise to interesting and experimentally accessible immunological questions.

To do this, we will first discuss the remarkable and rapid adaptive evolution which occurs during the immune response. The last term refers to the process by which the immune system, in response to a specific antigen, "tunes" the antibody molecules that it secretes. These antibody molecules accumulate successive mutations which progressively increase their affinity for the incoming antigen.

THE NATURE OF THE IMMUNE RESPONSE

When an organism such as a human is exposed to an antigen and mounts an immune response, the complex sequence of events which ensues includes binding of the antigen to immature antibody-bearing B cells. Those B cells whose antigen receptors, each with the same specificity antibody molecule that it will later secrete, best match the incoming antigen and proliferate most rapidly. This process is called *clonal selection*,[8] and leads to an abundance of antibodies in the blood serum which match the antigen.

The antigen specificity of an antibody immunoglobulin is determined by the amino acid sequence of its heavy- (H) and light- (L) chain variable regions. The variable region's diversity is generated by the combinatorial assembly of five different variable (V) gene segments by genomic rearrangement events during the formation of the particular complete V genes, which are between 330 and 360 base pairs long. A complete heavy-chain V domain results from the joining V_H, diversity (D), and H-chain joining (J_H) gene segments in the genomic rearrangements in each particular stem cell. Similarly, the L and V domain is created by joining of V_L and J_L gene segments. Each of these gene segments is chosen from a repertoire of several to several hundred alternatives, to build up combinatorially a very large number of alternative heavy-variable and light-variable regions.[29,63] Honjo[29] estimates the minimal diversity in the mouse generated by these mechanisms to be $5.1 \cdot 10^7$, whereas Berek et al.[5] estimate the diversity at 10^9.

In addition to this combinatorial diversity, two other sources of diversity are generated by variability in the exact locations of joining at the junctions of V gene segments during assembly with insertion of random nucleotides.[58] In addition another process results in nucleotide replacement, and is termed *somatic mutation*. In principle, such somatic mutation allows almost limitless V-region diversity. From analysis of clonally related cells, it now appears that there exists a special hypermutation system which specifically alters bases in the V region at a rate of 10^{-3} per base pair per generation, a rate approximately six orders of magnitude higher than the spontaneous mutation rate.[9,38,41,48,62]

We also note that, within either the heavy- or light-chain V region are three special sub-regions called *complementarity determining regions* (CDRs). These complementarity determining regions are thought to be the parts of the V region that actually bind the antigen. Presumably, the remaining amino acids, the *framework*, provide a superstructure for the CDR amino acids.

ANTIGEN SELECTION THEORIES

The cellular and molecular mechanisms by which the maturation of the immune response occurs are still being uncovered. Classical theories suggest that competition for limited amounts of antigen may drive a selection process.[55] The argument goes as follows: the amount of antigen bound to cell surface immunoglobulin depends upon the product of the antigen concentration and the affinity of the receptor for antigen. During the course of an immune response the antigen concentration should decrease. If there is a critical amount of bound antigen required to stimulate a B cell into antibody production, then as antigen concentration falls only those B cells with increasing affinities for the antigen will remain stimulated and continue to secrete antibody. Because the antibody secreted by a cell has the affinity for the antigen as its receptor, one should see the average affinity of serum antibody increase during the course of an immune response. According to this theory, based on clonal selection, smaller and smaller subsets of pre-existing B cells are selected by antigen during the immune response. Mathematical models based on this theory were developed by Bell.[3,4]

SOMATIC MUTATION THEORIES

Doria[13] pointed out that certain patterns of actually observed affinity changes are not consistent with the classical theory. For example, antigen selection theories would predict that low doses of antigen would lead to antibodies of higher affinity than would be found after giving higher doses. Instead, lower doses lead to antibodies with lower affinity.[54]

Recent evidence has been obtained by studying the messenger RNA (mRNA) sequences of monoclonal antibodies. Sequencing of mRNAs from different stages of the immune response to a single antigen indicate a more complex process. In particular, it now appears that somatic mutation plays a major role in maturation of the immune response such that the affinities of the secreted antibodies increase over time.

In response to a specific antigen, clonal proliferation of those germline genes whose variable regions most precisely match the antigen leads to amplification in the serum of an initial set of "roughed in" antibodies from a restricted number of V-region-containing cells. The initial fraction of B cells which responds to an antigen[45] is on the order of 10^{-5}. These germline genes have little or no somatic mutation evident.[30,38,58,62] Later in the primary or secondary response the majority of antibodies no longer directly correspond to germline varieties, but show extensive somatic point mutations. The accumulation of these point mutations is correlated with an increase in the affinity of the antibody for the antigen. According to present somatic mutation theories, the increased affinity is itself a direct consequence of further clonal selection. Those somatic mutations which result in an alteration of the protein sequence of the V region may alter the binding affinity of the antibody molecule for the antigenic determinant. Then those mutated B cells whose antibodies bind the antigen with higher affinity proliferate more rapidly, and come to dominate the immune response by clonal selection.

Over a succession of somatic mutations to the V region of the initial germ-line B cells, the mean affinity of the antibodies increases sharply. Typical changes in affinity over the course of maturation[20,30] are increased from 5×10^{-4} M to 5×10^{-7} M.

Maturation of the immune response is therefore an adaptive walk in antibody space from the initial roughed-in crudely matching germline recombinant V region amplified by clonal selection through a succession of higher affinity variants, to or towards some local optimum antibody which is of higher affinity than its mutant neighbors. All the questions we have posed previously regarding the character of adaptive walks come to the fore, and point to a central experimental question: how correlated is the landscape?

We define the "landscape" in question precisely. Consider the incoming antigen and a single epitope, or molecular feature on the antigen. Then consider measuring the affinity of all possible antibody molecules for the single epitope. The distribution of the affinity values across antibody sequence space constitutes a well-defined affinity landscape with respect to that specific epitope. Presumably it is the statistical character of that landscape which largely determines the character of adaptive walks in antibody space. Therefore, we would hope that studies of a model like the *NK* model might provide this kind of statistical information. We discuss a method of applying the NK model to explore these issues next.

APPLICATION OF THE NK MODEL TO THE MATURATION OF THE IMMUNE RESPONSE

Our fundamental assumptions in applying the NK model to the maturation of the immune response are that a representative member of the population of maturing antibodies can be identified at any time and that these antibodies steadily increase in fitness due to fortuitous point mutations until a locally optimal antibody is obtained. We define a locally optimal antibody to have higher affinity for the antigen than any of its 1-mutant neighbors. The experimental results in the preceding section confirm that this is, in general, a plausible scenario. However, it is not known whether mature antibodies are, in fact, local optima. It is known that the V regions continue to mutate without substantial changes in affinity even after they have attained maximum affinity for the antigen. This may reflect mutational dispersal among near neutral mutants in the immediate vicinity of the local optimum. We list our more detailed assumptions below.

1. *Choice of the parameter N.* We identify the parameter N with the number of amino acid sites in the V region. As indicated in the previous discussion, there are between 110 and 120 amino acids in a typical V region. We use $N = 112$ because it was slightly easier to simulate a chain whose length is a multiple of eight.

2. *Choice of the starting place.* We assume that the fact that one in one hundred thousand B cells responds to a given antigen implies that those that do respond secrete antibodies that are in the $99 \cdot 999$th percentile in ability to bind to the antigen. Walks start well up on adaptive hillsides. From the point of view of our simulations, the fitness contribution of each amino acid in the model antibody is a random number. Therefore, finding antibodies in the appropriate percentile was reduced to the problem of finding random number seeds that gives a sequence of $N = 112$ random numbers whose average is in the same percentile. Although our use of this procedure implies that there will be some fluctuation in the starting fitness of the model antibody, departure from the bottom boundary of this top percentile was insignificant.

3. *Choice of neighborhoods.* Preliminary simulations reported above suggest that the lengths of walks to optima and the fitnesses of the optima achieved do not depend strongly on the details of which sites interact with each other. However, those preliminary simulations assumed that there could only be two amino acids per site and that the walks started from randomly selected initial peptides rather than peptides that are already in the $99 \cdot 999$th percentile in fitness. In modeling V regions we again consider both extremes: each amino acid only interacts directly with its K neighbors; to avoid "boundary" effects we therefore idealized the V region as a circular protein. Alternatively, each amino acid interacts with K amino acids drawn randomly from the chain. In this case we do not assume the peptide chain is a circle.

4. *Choice of 19N neighbors or neighbors via the genetic code.* A V region of length N can be thought of as having $19N$ 1-mutant neighbors. However, at the DNA level, many single amino-acid substitutions require two or three base-pair changes. Restriction to single base changes at the DNA level implies a reduction on the number of 1-mutant neighbors at the protein level. Both cases were studied. We incorporated "coding" into the model by explicitly including translation. In particular, we modeled this by assuming that the evolving entity was a pair of polymers, a "protein molecule," consisting of the 112 "amino acids," as before, and a (single-stranded) "DNA molecule" consisting of $112 \times 3 = 336$ sites, each with one of the four "bases." The initial DNA molecule was "back-translated" from the starting model V region to a DNA sequence coding for that model V region. The condon assigned to each position in the back-translation DNA was chosen randomly from the synonymous condons for that amino acid. A "step" consisted in a point mutation of one of the DNA sites, translating the new DNA sequence into the corresponding protein using the genetic code, and then computing the fitness of the protein. Since adaptation passes only to *fitter* neighbors, in this procedure the adaptive walk does not pass to a 1-mutant neighbor which is a silent mutation to a synonymous condon. A DNA mutation

which resulted in an internal stop condon in the model V region was scored as a lethal mutation with fitness 0.

Use of the "code" sharply reduces the number of 1-mutant neighbors. Each DNA sequence has only 1008 1-mutant neighbors, which are obtained by substituting any of the three other possible bases in each of the 336 sites. In addition, due to synonymous codons, only about 75% of these result in substitution of a new amino acid. Thus in the versions of the model based on coding, each model V region has about 756 1-mutant neighbors rather than $19 \times 112 = 2128$ 1-mutant neighbors.

5. *Complementarity-determining regions (CDRs) or not.* The NK model, in its general form, is isotropic. It assumes that all sites make direct contribution to fitness of the overall string, whether that string is interpreted as a genotype or protein. Proteins, however, may be more hierarchically constructed, with some sites, e.g., amino acids in the actual active site of an enzyme or binding site of an antibody molecule, having direct bearing on function, while others play a more modest support role.

As remarked above, in the V region of antibody molecules, special hypervariable regions called complementarity-determining regions, are known to play a critical role in antibody diversity and in actual binding of the antigen. The surrounding parts of the V region are thought to be a supporting "framework" for the basic structure of the binding site. A simple way to begin to model the distinction between CDRs and framework is to assume that only the amino acids in the CDRs make direct impacts on the fitness of the V region, while those in the framework act via their influence on the CDR amino acids. Thus as a first effort we have modeled the existence of three CDRs by assigning three contiguous regions of amino-acid positions in our model V regions, matching those in observed V regions of these sizes and spacing, with a total of 37 amino acids, in our modeled V regions. Because the amino acids in the framework interact with the amino acids in the CDRs, they still have an indirect bearing on fitness.

6. *Choice of K.* At this point, the remaining free parameter in the model is K. The experimental data we describe in more detail immediately below show that walk lengths in actual affinity landscapes average between six and eight steps, but with considerable variance. Walks start well up on adaptive hill sides where the starting germline V region initially amplified by clonal selection is in the highest $99 \cdot 99$th percentile. Thus we seek a value of K such that walks to local optima from that starting percentile average six to eight steps. This is the central parameter matching step in applying the NK model. We use two features of the immunological data: the fraction of B cells which respond to an antigen sets the starting percentile in affinity space; the number of mutations substituting amino acids in the V region during maturation sets the mean walk length to optima. Given these we can *find* the value of K which yields walks with the appropriate length by carrying out numerical simulations at various trial values of K.

AFFINITY LANDSCAPES ARE CORRELATED. An immediately interesting point arising from framing these questions is that the appropriate value of K must be less
than the maximum, $K = N - 1$, and therefore that antibody affinity landscapes
must be *correlated*. This follows from examining walk lengths at the upper extreme
value for $K, K = N - 1$, corresponding to a fully random landscape. Here the probability that a model V region with rank-order fitness x is fitter than its $19N$ fitter
neighbors is x^{19N}. Thus, any starting peptide that is in the top $99 \cdot 999$th percentile
in fitness has roughly a 98% chance of already being fitter than its 2128 1-mutant
neighbors. That is, *if affinity landscapes were entirely uncorrelated, initially selected
germline variants would already be local optima.* Since maturation of such antibody
molecules does occur, we can conclude both that affinity landscapes are correlated,
and that, within the NK model, K must be less than $N - 1$.

EXPERIMENTAL FEATURES OF AFFINITY LANDSCAPES RAISED BY THIS ANALYSIS

Maturation of the immune response occurs on a rugged affinity landscape whose
structure is only partially known. Regardless of whether the NK model itself proves
to be a good model for the structure of that landscape, a major purpose in this
article is to focus attention on the structure of such landscapes. In general, all the
questions raised previously regarding abstract landscapes are *a fortiori* of interest
with regard to the immune system:

1. How many improvement steps must be taken from any initial antibody molecule
   to a local optimum, i.e., how many somatic mutations accumulate in the V
   region of an initial roughed-in germline variant antibody molecule during maturation? The answer, as mentioned above, appears to be a range, with a mean
   of six to eight.[5,7,28,58] For example, Crews et al.[11] studying the V_H gene responding to phosphoryl choline found between one and eight residues changed;
   Bothwell et al.[7] found three mutations in a lambda (λ) light-chain and six in a
   λ_2 light-chain V region; McKean et. al.[41] studying the V_k region of antibodies
   against a determinant on influenza found seven or eight replacements; Clark
   et. al.[9] studying the secondary response to influenza found 20, 12, and 19 V_h
   coding mutations, and 9, 8, and 15 V_k coding mutations.
2. What fraction of 1-mutant variants of the initial roughed in germline antibody
   have higher affinity for the immunizing antigen? How does that fraction change,
   presumably dwindling to 0, as successively higher-affinity antibody molecules
   are selected during maturation of the immune response? Here it is known that a
   large fraction of the 1-mutant variants have lower affinity, but the exact fraction
   with higher affinity at any step of the maturation processes is unknown.
3. How rugged is the affinity landscape in the 1-mutant vicinity of local optima?
   The question of whether affinity falls off dramatically in some directions and
   slowly in others translates directly to whether mutations at specific positions
   in the V region cause dramatic loss of affinity while those at other positions
   cause little loss of affinity. Restated, the distribution of the number of amino

acids which can be substituted at a site with retention of function is a direct picture of the local ruggedness of the affinity landscape.

4. How many alternative local optima can be reached from any initial roughed-in germline antibody amplified by initial clonal selection? Further, what is the probability of climbing to each of those alternative optima, hence the density of their occupancy? Here recent work with inbred mice[44,56] has demonstrated that multiple local optima are accessible. In many cases initial clonal selection opts for the same initial V region, which then climbs to different mature forms by accumulating different somatic mutations. It appears from these and similar experiments that the number of alternative optima accessible from the initial antibody may be at least modestly large. Typically, comparison of five to ten monoclonal antibodies deriving from the same initial V gene shows that all differ from one another. Because only small numbers of sequences have been compared in this way, it is unknown whether a much lager number of local optima are accessible.

These experiments are not entirely unambiguous. As remarked above, we have assumed that mature antibodies are actually local optima, and one of the predictions of the NK model will be that many local optima should be accessible. However, the fact that different mature antibodies emerge from the same V gene is insufficient to confirm our reasoning. From the work of Eigen and Schuster,[17] Eigen,[15] Schuster,[50] and Kauffman,[33,34] as well as classical population genetic analysis,[18] we know that a mutant spectrum around an optimum can be expected, and we know that the rate of hypermutation is high. Thus the diversity seen in mature antibodies derived from one initial V gene may reflect the incapacity of clonal selection to eliminate near neutral variants.

5. How similar are the local optima? Maturation climbs to alternative local optima from an initial roughed-in V region. Described more fully below, the typical observations, when several different monoclonal antibodies derived by maturation are compared is that many amino acids are "conserved" while a smaller fraction are repeatedly mutated with respect to the initial V gene. Furthermore, some sites repeatedly have mutated "in parallel" to the same alternative amino acid.[56]

As stressed, we use observed walk lengths and starting percentile to "tune" K. Hence fitting walk lengths with the NK model is just curve fitting. However, the value of K which we drive, and the remaining features (2)–(5), are aspects of the immune response about which the NK model makes clear and testable predictions. We return to this below after describing our simulation results.

PREDICTIONS OF THE MODEL AND COMPARISON WITH EXPERIMENTAL DATA
THE APPROPRIATE VALUE OF K IS NEAR 40

Numerical simulations were carried out for all versions of V-region models. Of these, presumably the most realistic combination includes both the CDRs and the genetic code. But as we will see, all of the possibilities predict the same features of the landscape and qualitatively agree with the available experimental data. The results are remarkably robust. As Table 4 shows, whether CDRs were included or not, whether all $19N$ protein neighbors of the V gene were used or translation via the genetic code were used, and whether the K sites were constrained to be flanking adjacent sites or choosen at random, a value of $K = 40$ gives rise to walks of between 6 and 12 steps. $K = 30$ typically yields walks which are too long. $K = 50$ typically yields walks which are too short. Since walk lengths are largely insensitive to the remaining parameters, to a very good first approximation, the dominant parameters are N and K.

There is considerable dispersion about this mean value. As shown in Figures 4(a) and (b), for a given value of N and K and under defined conditions for the rest of the model conditions, walk lengths might range from 2 or 3 to 15 to 20. This dispersion reflects again the ruggedness of the landscape and is encouraging, given the fact that there is a similar dispersion in the experimental data.

Ultimately, the NK model predicts some specific *distribution* of walk lengths to optima, not just a mean and a standard deviation. Thus accumulation of adequate experimental data can ultimately establish the actual observed distribution of walk lengths for comparison to the NK model or improved models.

Finding a specific value for K is, in itself, interesting. If the model is taken literally, K stands for the number of amino acids which bear on the fitness contribution of each amino acid. Then, if K is about 40, alteration in a single amino acid could affect the behavior and function of roughly 40 amino acids in the V region. Is this plausible and is there any evidence bearing on the issue?

In a well-folded protein, an amino acid is not only open to influence by those which are its neighbors in the primary sequence, but also those which are near it in the folded form due to juxtaposition with amino acids distant in the primary sequence. For example, each of the CDR loops is typically hydrogen bonded to one or more of the other CDRs as well as points in the framework.[52] In turn, any of those amino acids is itself coupled to its neighbors in the primary sequence.

One approach to studying this question is based on hydrogen exchange data. The experimental technique consists of observing a protein in deuterated water. Hydrogen atoms on different amino acids in the protein exchange with the deuterium at different rates which depend upon the product of an intrinsic exchange rate for a hydrogen exposed to deuterated water and frequency of exposure. That

TABLE 4 Mean walk length to local optima, mean number of fitter neighbors from the first model V region on the walk, and mean number of sites of the locally optimal V region attained which could be substituted with maintenance of above "threshold" fitness. Simulations were carried out for different versions of the model, as shown. The number of trials under each condition was ten with conditions using the DNA code and 25 for those without it. Standard deviations shown in brackets.

	K	Average walk length	Average no. fitter neighbor on 1st step	Mean no. of allowed subst./site
Protein (no cdr)				
adjacent	30	$13 \cdot 6(3 \cdot 2)$	$65 \cdot 9(20 \cdot 4)$	$15 \cdot 4(3 \cdot 6)$
	40	$8 \cdot 6(4 \cdot 5)$	$24 \cdot 7(9 \cdot 9)$	$4 \cdot 1(3 \cdot 9)$
	50	$4 \cdot 9(2 \cdot 3)$	$10 \cdot 3(4 \cdot 9)$	$0 \cdot 8(1 \cdot 4)$
random	30	$17 \cdot 7(4 \cdot 4)$	$83 \cdot 7(13 \cdot 0)$	$17 \cdot 9(2 \cdot 4)$
	40	$11 \cdot 5(5 \cdot 3)$	$42 \cdot 1(6 \cdot 9)$	$8 \cdot 5(5 \cdot 8)$
	50	$6 \cdot 6(2 \cdot 2)$	$17 \cdot 5(3 \cdot 5)$	$2 \cdot 1(3 \cdot 0)$
Protein (with cdr)				
adjacent	30	$26 \cdot 1(7 \cdot 1)$	$89 \cdot 7(27 \cdot 8)$	$12 \cdot 5(6 \cdot 6)$
	40	$9 \cdot 8(3 \cdot 7)$	$27 \cdot 7(11 \cdot 3)$	$5 \cdot 0(5 \cdot 5)$
	50	$7 \cdot 0(3 \cdot 3)$	$12 \cdot 7(6 \cdot 0)$	$1 \cdot 5(2 \cdot 9)$
random	30	$16 \cdot 1(5 \cdot 3)$	$81 \cdot 8(16 \cdot 2)$	$15 \cdot 5(4 \cdot 5)$
	40	$12 \cdot 0(3 \cdot 9)$	$36 \cdot 9(9 \cdot 3)$	$7 \cdot 4(5 \cdot 8)$
	50	$7 \cdot 3(3 \cdot 2)$	$17 \cdot 0(5 \cdot 8)$	$1 \cdot 9(3 \cdot 0)$
DNA Code (no cdr)				
adjacent	30	$8 \cdot 4(2 \cdot 3)$	$20 \cdot 1(12 \cdot 7)$	$10 \cdot 5(5 \cdot 1)$
	40	$5 \cdot 2(2 \cdot 7)$	$7 \cdot 3(4 \cdot 3)$	$3 \cdot 4(3 \cdot 8)$
	50	$2 \cdot 9(2 \cdot 1)$	$3 \cdot 9(2 \cdot 5)$	$0 \cdot 3(0 \cdot 7)$
random	30	$11 \cdot 7(4 \cdot 5)$	$24 \cdot 6(8 \cdot 2)$	$14 \cdot 1(5 \cdot 1)$
	40	$6 \cdot 9(3 \cdot 3)$	$11 \cdot 2(4 \cdot 9)$	$5 \cdot 7(5 \cdot 1)$
	50	$3 \cdot 1(1 \cdot 9)$	$4 \cdot 5(2 \cdot 8)$	$1 \cdot 0(2 \cdot 0)$
DNA Code (with cdr)				
adjacent	30	$15 \cdot 1(5 \cdot 6)$	$41 \cdot 8(17 \cdot 5)$	$9 \cdot 7(7 \cdot 5)$
	40	$6 \cdot 7(3 \cdot 1)$	$17 \cdot 1(12 \cdot 1)$	$3 \cdot 2(4 \cdot 6)$
	50	$3 \cdot 5(2 \cdot 1)$	$7 \cdot 0(7 \cdot 0)$	$0 \cdot 6(1 \cdot 7)$
random	30	$11 \cdot 2(4 \cdot 1)$	$27 \cdot 4(9 \cdot 3)$	$11 \cdot 9(6 \cdot 1)$
	40	$7 \cdot 6(2 \cdot 7)$	$10 \cdot 6(5 \cdot 5)$	$3 \cdot 8(4 \cdot 5)$
	50	$4 \cdot 1(2 \cdot 2)$	$5 \cdot 6(3 \cdot 5)$	$0 \cdot 7(1 \cdot 7)$

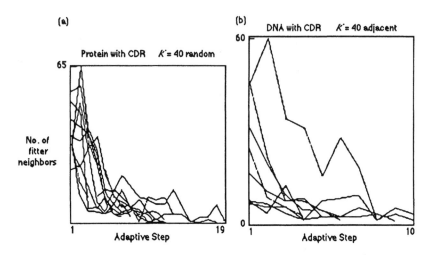

FIGURE 4 The number of fitter 1-mutant neighbors at each adaptive step from the initial model V region, the best in 100,000, on several different adaptive walks to local optima for the same values of the NK model parameters. (a) Modeling the V region with CDR regions present, and with $19N$ 1-mutant neighboring proteins. (b) As in (a) except that 1-mutant neighbors were determined via the DNA code.

frequency in turn depends upon subtle "breathing" motions of the protein as it twists and unfolds slightly in different ways. Thus hydrogen exchange is a sensitive measure of protein behavior. Wand et. al.[59] and Roder et. al.[46] have studied such hydrogen exchange in numbers of proteins. While they have not yet analyzed differences between a protein and a 1-mutant variant, they have looked at the oxidized and reduced form of cytochrome C, a 106-amino-acid protein. Oxidization and reduction, due to the presence or absence of a charge on the heme group, correspond very roughly to substitution of a charge for an uncharged amino acid in that vicinity. Roder et. al.[47] have been able to examine 50 hydrogen-bonded pairs of amino acids, and find that at least 30 of them alter their exchange behavior in the oxidized and reduced forms. The very crude conclusion to be drawn is that a charge alteration at one point in a protein can affect at least 30 amino acids. Since these authors studied only half the hydrogen-bonded atoms, the number of amino acids affected by altering one amino acid may be greater than 30. This point has obvious caveats. The study is not of an amino-acid substitution, but of an altered heme group. Further, to have found a statistically significant alteration in hydrogen exchange by an amino acid does not yet say that such alterations are in any way relevant to the function of the protein. Third, the cytochrome C molecule may be well evolved to undergo alterations when the heme group is charged. Many fewer alterations in hydrogen exchange behavior might be found by randomly substituting amino acids in proteins. Nevertheless, the data suggests that any amino acid might be affected by, and affect, as many as 30 amino acids in a protein region of about 106 amino acids. Direct testing in antibody molecules would require study of hydrogen exchange in the V region of a mature antibody and its 1-mutant variants.

THE NK MODEL MAKES PLAUSIBLE PREDICTIONS ABOUT THE FRACTION OF FITTER 1-MUTANT VARIANTS

Given a value of $K = 40$, the NK model makes clear predictions of the first roughed-in V region and of each improved variant on the adaptive walk. The expected number of fitter variants to the first V region is on the order of one or two percent in all of the combinations of conditions mentioned in Table 4. In those runs that used all $19N$ 1-mutant neighbors, 24–42 among the 2128 1-mutant variants are typically fitter. When translation via the genetic code (and the implicit constraints in the 1-mutant neighbors) were added, typically there are about 7–17 fitter 1-mutant variants among the 1008 1-mutant nucleotide substitutions and about 756 1-mutant V regions at the protein level (Table 4).

There is moderate variance in the fraction of fitter 1-mutant variants of the initial V region on individual walks. The minimum we have found on the initial step is 1, and the maximum is 70, or over 3%.

The fraction of fitter variants dwindles, but not smoothly on any specific walk, to 0 over the steps to the local optimum, as shown in Figure 4(a) and (b).

The actual fraction of fitter variants in maturing antibody molecules is not yet known in detail, but the experimental procedure to find this fraction is clear: Monoclonal antibodies at different stages during an adaptive walk must be obtained, the gene cloned, and the 1-mutant spectrum examined for the affinities of the 1-mutant variants. However, studies of the *lac* repressor provide an indirect estimate of the number and the nature of fitter 1-mutant variants of a roughed-in V region.

The *lac* repressor monomeric unit has 360 amino acids. Miller et al.[52] studied a collection of over 300 altered proteins, each by a single substitution, with respect to both ligand-binding activities. The mutant forms were generated in a controlled way by use of 90 different nonsense sites in the corresponding *lac* I gene which were suppressed by a set of nonsense suppressors of amber (UAG), ochre (UAA), or UGA mutations. The classes of substitutions allow substitution of similar (e.g., hydrophobic for hydrophobic) and dissimilar amino acids at 25% of the positions in the normal molecule.

Mutant phenotypes due to loss in capacity of the repressor molecule to bind to the operator DNA, or to bind allo-lactose (or the synthetic inducer IPTG), or due to an increased affinity for the DNA or IPTG, allowed Miller's group to study the consequences of such mutations conveniently. The overall results from 323 single amino-acid replacements is about 42% result in a detectable change in either the capacity to bind IPTG, or the operator DNA. The remaining 58% appear to be "silent" and do not result in measurably altered proteins. About 33% of the replacements decrease capacity to bind to the operator DNA, although only 15% of the substitutions destroyed 25% or more of the capacity, and only 8% became fully inactive. About 11% reduced affinity for the inducing metabolite. On the other hand, 1% of all one-step mutants actually increased affinity for the operator DNA, in some cases by as much as 100-fold. No one-step mutant was found which increased affinity for IPTG.

From this study we can draw at least the following conclusions. First, even well-tuned proteins may have rare variants which "improve" a given function. Here, one percent of the 1-mutant neighbors, at a restricted number of sites, showed increased affinity for the operator DNA, and no mutant was found which showed increased affinity for the inducer metabolite. Second, 58% of the single amino-acid substitutions had no obvious effect. Because the assays employed are rough measures, the reported fraction of neutral mutations is probably an overestimate. However, since 42% of the mutants clearly do reduce affinity for the operator or inducer or both, it is very unlikely that more than a small fraction of the one-step neighbors subtly increase affinity.

Our simulation results suggest that an initial germline V region with $K = 40$ would be open to improvement by about 1–2% of the 1-mutant variants. This is very close to the observed data for the *lac* repressor. Thus, tuning K to fit observed walk lengths yields a value which, having tuned the ruggedness of the fitness landscape, predicts a plausible value for the expected fraction of fitter 1-mutant variants of the initial germline V region amplified by clonal selection.

Note that these predictions of the NK model are fairly sensitive to K. When $K = 30$, roughly 3–5% of the 1-mutant variants of the first clonally selected V region have higher affinity, while walks to optima would average about 13 steps. For $K = 20$, the average walk length is 22 steps and about 7% of the 1-mutant variants of the initial antibody are fitter.

THE NK MODEL PREDICTS THE EXISTENCE OF CONSERVED AND VARIABLE SITES IN THE V REGION

In real proteins, antibodies and otherwise, it is widely known that some amino-acids cannot be altered without drastic loss of function while amino acids at other positions can be altered with relative impunity. It is therefore of interest to ask whether the NK model, for the parameters given, predicts this phenomenon without further assumptions.

To answer this question, we computed the fitnesses of all of the 1-mutant neighbors of the local maxima obtained in our walk simulations. In order to make valid comparisons between simulations with and without the genetic code, we substituted all 19 other amino acids in each site of the V region, including those amino acids that required several mutations of the corresponding DNA sequence. The results, for different values of K and for different assumptions about whether the entire V region or just the CDRs are used in computing fitness, are as expected. The first main result is that as K increases, the jaggedness of the landscape increases. In other words, landscapes are smoother for low K than for high K. The second major result for $K = 40$ is perhaps more surprising: at some sites, any model amino-acid substitution causes a dramatic loss of fitness, while at others, all substitutions cause almost no loss of function. At still other sites, some substitutions cause almost no loss of function while other amino acids in the same site cause drastic loss of function. Thus, without further assumptions, the NK model for these parameters gives a highly rugged landscape in which amino acids at some sites in the locally optimal V region must be entirely conserved to preserve function, and amino acids at other sites can be substituted indiscriminately.

Note that, in constructing the general NK model, no site is *a priori* more important than others. It is instead the fact that K is high, resulting in a rugged landscape which predicts that some sites are conserved and others broadly substitutable.

A particularly interesting view of these results is the following. We have no direct scaling relating "fitness" in the NK model with real affinities of antibody molecules. However, real antibody walks start with those already the best in 100,000, and such antibody molecules typically have affinities of about 10^{-4} M for their antigens. In contrast, matured antibodies have affinities around 10^{-7} M. Then it is sensible to define the fitness of the first member of the model walk as corresponding to a modest affinity of 10^{-4} M, and let this fitness serve as a threshold separating model V regions which do and which do not bind the antigen. Given this threshold, we can test the number of substituted amino acids at each site in the 112-long optimal V region which preserve "at least above-threshold" function. Figures 5(a) and (b) show the results for model V regions for different values of K. Similar results are found for the different versions of the model with and without CDR, coding, or choice, or adjacent or random epistatic connections (Table 4). Again, it is K that determines the qualitative features of the landscape. For $K = 20$, each of the 112 sites can be substituted by all 19 of the other amino acid and the mutated model V region remains above-threshold in affinity. For $K = 30$, most sites are substitutable by 19 amino acids, but some sites can be substituted with fewer amino acids and preserve above-threshold affinity. But for K about 40, a very wide distribution is found. Some sites can be substituted by 19 other amino acids, some by 15, some by 10, some by 5, and some by 0. Thus K about 40 yields the broadest distribution. We emphasize that this broad distribution is a *prediction* of the NK model.

The experimental data to test this prediction would consist of a high-affinity mature monoclonal antibody against a defined epitope, and its entire 1-mutant spectrum with respect to V-region mutants. The affinities of the mutant spectrum constitute the data set. It is not available, but the experiment is obviously feasible using cloned antibody molecules. Nevertheless a rough approximation to this experiment is available. Geysen et al.,[23,24,25] Getsoff et al.,[22] and Fieser et al.[19] have studied the effects of all possible 1-mutant variations in an *antigen* upon the antigen's affinity for the antibody. More precisely, in these studies, the authors raised polyclonal sera or monoclonal antibodies against a defined six-amino-acid-long epitope

TABLE 5 (a) The distribution of the number of sites in a locally optimal model V region with 112 sites which are substitutable by 0, 1, ..., 19 other amino acids with maintenance of "affinity" or fitness above a "threshold level" defined by the affinity or fitness at the start of the adaptive walk. $K = 20$ or 30 adjacent.

Allowed substitutions per site	$K = 20$		$K = 30$		
0	0	0	0	0	
1	0	0	0	0	
2	0	1	0	0	
3	0	1	0	0	
4	0	2	0	0	
5	0	2	0	0	
6	0	3	0	0	
7	0	5	0	0	
8	0	4	0	0	
9	0	20	0	1	
10	0	7	0	0	
11	0	18	1	5	
12	0	10	5	6	
13	0	11	6	3	
14	0	8	4	7	
15	0	10	7	12	
16	0	5	12	16	
17	0	4	12	14	
18	0	1	21	25	
19	112	0	44	23	
	19.00	11.13	17.14	16.37	Mean
	0.00	3.27	2.17	2.39	S.D.

on a protein antigen, then made synthetic hexamers identical to that epitope and demonstrated that the hexamer was bound by the sera or monoclonal antibody at high affinity. Then in each case the authors looked at all 20 variants at each of the six positions, one position at a time. The results for nine such peptides are summarized in Figure 6. In this summary we have utilized the authors, data and set an arbitrary threshold of about 10% affinity compared to the "wild-type" hexamer as the criterion for "function." The striking feature is that the distribution is again very broad. Some sites are not substitutable at all, others are subsititutable by all 19 other amino acids, and still other sites accept some fraction of the 19 and retain affinity.

TABLE 5 (b) As in (a), except $K = 40$ or 50.

Allowed substitutions per site	$K = 40$			$K = 50$			
0	7	3	7	105	63	85	
1	4	4	14	6	31	20	
2	11	5	6	1	13	6	
3	10	14	8	0	4	1	
4	15	9	3	0	0	0	
5	15	9	5	0	0	0	
6	12	13	5	0	0	0	
7	11	9	3	0	0	0	
8	7	6	5	0	0	0	
9	6	8	6	0	0	0	
10	3	5	6	0	0	0	
11	5	6	5	0	0	0	
12	2	4	7	0	0	0	
13	2	5	6	0	0	0	
14	1	5	4	0	0	0	
15	0	2	9	0	0	0	
16	1	4	8	0	0	0	
17	0	0	3	0	0	0	
18	0	0	2	0	0	0	
19	0	1	0	0	0	0	
	5.49	7.23	8.17	0.07	0.65	0.31	Mean
	3.36	4.29	5.62	0.29	0.88	0.61	S.D.

An interesting feature of the observed distribution is that it is not only broad, but bimodal. Sites are more likely to be entirely substitutable, or not substitutable at all. A bimodal distribution emerges rather naturally from our model if a *small distribution of K values* between 30 and 40 is assumed. Whether the bimodality is to be taken seriously at this state is uncertain.

Four comments are warranted. First, it is clearly encouraging that the NK model predicts a broad distribution for K near 40, and such a broad distribution is found. Second, we have defined a "threshold" affinity as the fitness of the first model V region in the walk, the best in 100,000. We do not know how this threshold bears on the experimental affinities measured. Third, the experimental data need to be used cautiously in this context. They concern free hexamers bound to polyclonal

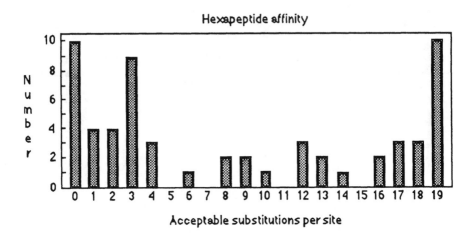

FIGURE 5 Experimental distribution of acceptable number of amino-acid substitutions per site in hexamers with maintenance of 10% or more of the affinity of the "wild-type" hexamer for the polyclonal sera or monoclonal antibody.

sera, or monoclonal antibodies, not the number of substitutions at each position within the V region of a mature monoclonal antibody molecule. The constraints within a V region may or may not dramatically alter the observed distribution. Fourth, taking data and model for the moment at face value, the same value of K which fits walk lengths to optima also predicts a reasonable fraction of fitter 1-mutant variants, and genuinely predicts that some sites allow no substitutions while others are more permissive. Were K much smaller, say 20, almost all sites would be open to substitution by most model amino acids. The prediction is thus sensitive to K.

THE NUMBER OF ALTERNATIVE LOCAL OPTIMA FOUND FROM AN INITIAL V REGION

The NK model for these parameters allows us to examine the number of alternative optima accessible from the initial model V region, and also to test whether alternative accessible optima are typically "hit" equally often on independent walks, or with biased preferences.

The experimental data on repeated walks from the same V region remain scant, as noted, but clearly suggest that multiple optima are accessible from the same initial V region. The true number of such local optima is not known experimentally, but presumably is greater than the five to ten alternatives often observed.

Numerical simulations with $K = 40$ were carried out from initial model V regions, based on use of the DNA code, and were stopped by limitations of computer storage. In two simulations making 797 and 315 such walks from the same initial V region, 150 and 235 optima were found. Because many of these optima were found only once, it is difficult to know how many more remain to be accessed from the same initial V region. However, it is a clear prediction of the NK model that a given initial germ-line V region can give rise to hundreds, perhaps thousands of mature antibodies, each a local optimum in affinity space.

A second feature of these studies is shown in Figure 7, which is the histogram of the numbers of times each local optimum was "hit" on independent random walks from the same initial model V region. As can be seen, four optima are each encountered about 130 times. Analysis of these four showed that each is a 1-mutant variant of the initial V region from which walks started. Ultimately, the NK model predicts a distribution which is open to experimental testing. The density distribution with which nearby local optima are reached is another expression of the ruggedness of the fitness landscape.

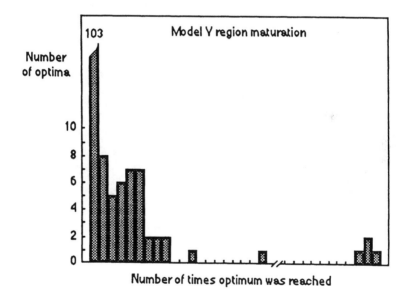

FIGURE 6   Multiple adaptive walks from the same initial model V region, $K = 40$, showing how many optima were "hit" once, twice, or many times. Note that some optima were encountered 130 times on independent walks from the same initial V region.

THE SIMILARITY OF ALTERNATIVE LOCAL OPTIMA: CONSERVED SITES AND PARALLEL MUTATIONS

Comparison of alternative mature V regions obtained experimentally reveals that not all sites in the V region accumulate somatic mutations equally. In particular,

TABLE 6 Comparison of numerical results for model V regions with different parameter values, and observed alternative V regions derived from a single initial V region, with respect to the deviation in the number of mutated and non-mutated sites from chance. In all cases, model and real V regions show that fewer sites are mutated than expected by chance, suggesting "hot spots" and conserved regions.

K	Number of observations		Number of experiments	Number of sequences
A 20	79	<	91	9
A 25	68	<	83	9
A 25	55	<	68	8
A 30	48	-	48	7
A 30	52	<	62	10
A 40	41	<	54	5
A 40	36	<	43	9
A 40	35	<	39	4
A 40	27	<	33	10
A 40	41	<	54	7
A 40	34	<	37	7
A 50	20	-	20	4
A 50	18	<	27	11
A 50	15	<	17	4
A 50	21	<	23	6
R 40	47	<	53	5
Antiarsonate vs "germ-line"	13	<	23	6
Antiarsonate vs prototype antiars	28	<	42	6
Antiphosphocholine vs "germ-line"	21	<	26	8

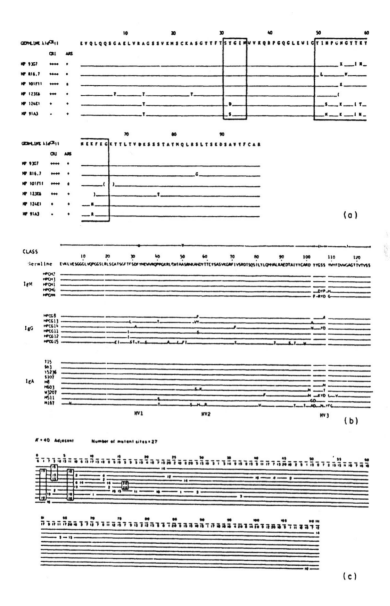

FIGURE 7 Observed initial antibody sequence early in the immune response and alternative mature forms which evolve from each. Letters on mature strands show substitutions relative to the initial germline V region from which maturation occurs, top strand. (a) The arsonate system, Ars-A.[56] Boxes correspond to the three CDR regions. (b) The anti-phosphocholine system.[44] (c) Comparison of initial model V region and ten different local optima found on adaptive walks from the initial V region. Numbers on different "mature" strands correspond to substitutions with respect to initial model V region. Boxes show parallel substitutions arising independently on independent adaptive walks to different local optima.

some sites are rarely mutated, and among the sites which are preferentially mutated, sometimes the *same* amino acid is substituted on two or more independent walks. These are called "parallel" mutations. To see whether these phenomena are observed in the NK model landscapes, we compared four to eleven alternative optima accessed from the same initial V region. Similarly, experimental data sets often compare five to seven V regions obtained by independent walks from the same initial V region.[44,56] Table 6 shows that for experimental and model V regions the numbers of sites which accumulate mutations in sets of the local optima compared to the initial V region is less than expected by chance. This means that some sites are preferentially not mutated, and others are mutated more often than expected in real and model V regions. Figures 7(a) and (b) show experimental data sets for two different clusters of V regions: one for the arsonate system[56] and other for phosphocholine.[44] In addition Figure 7(c) shows ten local optima and the initial model V region for an example with $K = 40$. Note that in the real and experimental sets some sites have similar parallel mutations.

CONCLUSIONS, CAVEATES, AND DIRECTIONS FOR FUTURE WORK

How seriously should we take the NK model as an account of the structure of affinity landscapes? With considerable, but not unbridled enthusiasm. The NK model is the first effort at a statistical model to predict the rugged structure of fitness landscapes in sequence space. A single choice of parameter values, $N = 112$ as set by the known length of the V region, and K about 40 as tuned to fit known walk lengths to mature antibodies predicts a number of features of antibody affinity landscapes well. It is premature to say that the NK model predicts these features accurately. All we can say now is that the predictions are very plausible. Direct experimental investigation with cloned V regions at different stages of maturation are needed to test the predictions. Even more directly, one might imagine carrying out an entire adaptive walk by fitter 1-mutant variants *in vitro* beginning with a cloned V gene from the initial B cells which respond.

Although broadly successful, the NK model as tested does exhibit certain failures. In particular, during maturation of the immune response there appears to be a tendency for mutations causing amino-acid substitutions to accumulate preferentially in the CDRs. Further, there may be a tendency for less than the expected number of mutations causing substitutions to accumulate in the framework regions outside the CDRs.[51] If true, these biases are not captured in our current application of the *NK* model to V regions. Such biases might reflect evolutionary specialization of the framework to create the fundamental structure of an antibody binding site, while the CDRs specialize for antigen binding. In this view, the framework is highly adapted and easily disrupted, leading to overall loss of binding by the entire V region. Modeling such a *high-adapted* character of the framework is ignored in our modeling of CDRs and frameworks. Instead we tested the case in which only CDRs make direct contributions to fitness and the framework acts indirectly via the CDRs. An alternative approach would be to allow the framework amino acids, on average to draw their K inputs largely from within the framework while CDR's draw their K inputs largely from within the CDR. This would allow separate evolutionary optimization of framework and CDR regions. Use of a spatially non random distribution of K values as epistatic "inputs" or "outputs" might provide a better model of the possible hierarchical epistatic relations among amino acids to overall functions.

The fact that the NK model appears to succeed as well as it does is encouraging in at least three respects. First, it suggests that a statistical model may well capture the actual structure of fitness landscapes. Second, if the NK model, or an improved model, can predict the statistical structure of antibody affinity landscapes, it may also be able to predict the structure of fitness landscapes with respect to enzymatic function. Both involve the evolution of a structure with a "business end"—the antigen binding site in the case of the antibody and the active site in the case of the protein. Third, if the NK model is close to right, it may be telling us something fundamental about how proteins work. In solid state physics, spin-glass models[6,57] capture the real behavior of physical spin glasses by assuming interactions are so complex that the statistical distribution of their effects can be captured by random assignments of coupling energies. The same may be true for proteins. We comment briefly on these issues.

Certainly the most important implications of the rough success of the NK model at this stage is the hint that some statistical theory may some day actually fit well-established data on the actual structure of the affinity landscape. Obvious refinements of the model would include more of the details of protein chemistry. Thus, as suggested above, some sites should have more interactions than others, reflecting the fact that some amino acids have more hydrogen bonds, hydrophobic bonding, and salt bonds than others. One would therefore like K to be chosen from a distribution of possible values. In this simplest application of the NK model the identity of the A amino acids at each site bears no relation to the identity at another site. Amino acid no. 7 is merely a name specific to each site. In reality, alanine at each site is the same amino acid. Thus, the nature of the interactions should reflect the fact that alanine at each site is the same amino acid and that different amino acids have different chemical properties.

Whether or not the theoretical predictions regarding adaptive walks on correlated landscapes generated by the NK model fit the experimental data is far less important than obtaining real insight into the true adaptive landscape of antibody molecules with respect to affinity for a specific epitope. It is worth stressing again the data needed to extend understanding. Perhaps most important is analysis of the affinity of *all* 1-mutant variants of a given antibody molecule for the same epitope. Second, as the immune response matures, and higher-affinity antibodies are amplified, it is important to examine whether the number of fitter V regions decreases as affinity increases. Third, we need to know the distribution of amino-acid substitutions which allow modest affinity for the epitope. Is is broad or not? Fourth, we need good data on sequence similarities during branching walks to alternative optima from the same V region. As noted, perhaps the best way to obtain the requisite data is to carry out *in vitro* adaptive walks from cloned examples of the initial germline variant amplified by clonal selection. At each step all 1-mutant variants should be generated, and one or more selected to carry on the walk to local optima. The actual structure of affinity landscapes is open to direct investigation and is one piece of the immune system puzzle.

We have focused in this article on the structure of affinity landscapes. But the immune response itself depends upon clonal selection of mutating B cells flowing across this landscape under the drives of antigen stimulation in the context of regulatory effects due to T cells, growth factors, and anti-idiotype effects, as well as the proliferation of T cells and other cellular components of the immune response. An obvious immediate direction for investigation is exploration of the proliferation and flow of B-cell populations across rugged affinity landscapes. As in models of molecular evolution on rugged landscapes,[60,5,6] a variety of behaviors including "freezing" of the adapting population into small regions of the space, and diffusive flow among near-neutral mutants along ridges in the affinity landscape, as functions of the mutation rate, population size, and fitness landscape structure, are to be expected.

ARE PROTEIN ADAPTIVE LANDSCAPES AND FOLDING LANDSCAPES RELATED?

We close with a question. Adaptive landscapes appear to be very rugged and may be captured by something like the NK model. The clear relation between the NK model and spin glasses was noted above. Spin-glass models are currently proving useful as models for protein folding itself,[61,14,59] and protein folding is a complex process of "self binding," rather than binding to another molecule. Spin-glass models stress the idea that the potential surface guiding protein folding is likely to be very complex with many local minima. Proteins, once folded, presumably "breathe" by undergoing transitions between these minima. Clothia[62] and Karplus et al.[14] comment that families of evolutionarily related proteins undergo shape deformations of their crystallized form on the same scale as the breathing deformations of a single protein. This suggests that the range of readily available shape deformations of proteins guided by intramolecular forces is closely related to the range of shape and function deformations in protein evolution. In turn, the function of proteins in binding ligands and catalyzing reactions is primarily due to the similar shape and force properties. Might it be the case that the statistical character of the potential surface underlying folding of proteins is intimately related to the statistical character of adaptive landscapes in protein evolution? If so, then spin-glass models, or the NK model or a similar improved model, may capture the right statistical features of both.

ACKNOWLEDGMENTS

This work was partially supported under ONR grant N00014-85-K-0258 and under NIH GM 40186.

REFERENCES

1. Anderson, P. W. "Spin Glass Hamiltonians: A Bridge Between Biology, Statistical Mechanics, and Computer Science." In *Emerging Synthesis in Science, Proceedings of the Founding Workshops of the Santa Fe Institute*, edited by D. Pines. Santa Fe Institute Studies in the Sciences of Complexity, Proc. Vol. I. Reading, MA: Addison-Wesley, 1987.
2. Ansari, A., J. Berendzen, S. F. Browne, H. Frauenfelder, I. E. T. Iben, T. B. Sauke, E. Shyamsunder, and R. D. Young. "Protein States and Protein-quakes." *Proc. Natl. Acad. Sci. USA* **82** (1985):5000–5004.
3. Bell, G. "Mathematical Model of Clonal Selection and Antibody Production." *J. Theor. Biol.* **29** (1970):191–232.
4. Bell, G. "Mathematical Model of Clonal Selection and Antibody Production. II." *J. Theor. Biol.* **33** (1971):339–378.
5. Berek, C., G. M. Griffiths, and C. Millstein. "Molecular Events During Maturation of the Immune Response to Oxazolone." *Nature (London)* **316** (1985):412–418.
6. Binder, K., and A. Young. "Spinglasses: Experimental Facts, Theoretical Concepts, and Open Questions." *Rev. Mod. Phys.* **54** (1986):801.
7. Bothwell, A. L. M., M. Paskind, M. Reth, T. Imanishi-Karl, K. Rajewsky, and D. Baltimore. "Somatic Mutation of Murine Immunoglobulin Lambda Light Chains." *Nature (London)* **298** (1982):380–382.
8. Burnet, M. *The Clonal Selection Theory of Acquired Immunity*. Cambridge, MA: Cambridge University Press, 1959.
9. Clark, S. H., K. Huppi, D. Ruezinsky, L. Stuadt, W. Gerhard, and M. Weigert. "Inter- and Intra-Clonal Diversity in the Antibody Response to Influenza Hemagglutinin." *J. Expl. Med.* **161** (1985):687–704.
10. Clothia, C., and A. M. Lesk. "The Evolution of Protein Structures." In *Cold Spring Harbor Symposia on Quantitative Biology*, vol. LII. New York, NY: Cold Spring Harbor Laboratory, 1987, 399–405.
11. Crews, S., J. Griffin, H. Huang, K. Calame, and L. Hood. "A Single V_h Gene Segment Encodes the Immune Response to Phosphorylcholine: Somatic Mutation is Correlated with the Class of the Antibody." *Cell* **25** (1981):59–66.
12. Derrida, B. "Random Energy Model: An Exactly Solvable Model of Disordered Systems." *Phys. Rev. B* **24** (1981):2613.

13. Doria, G. "Immunoregulatory Implications of Changes in Antibody Affinity." In *Regulation of Immune Response Dynamics*, vol. II, edited by C. DeLisi and J. R. J. Hiernaux. Boca Raton, FL: CRC Press, 1982.

14. Edwards, S. F., and P. W. Anderson. "Theory of Spin Glasses." *J. Phys. F: Metal Phys.* **5** (1975):965–974.

15. Eigen, M. "Macromolecular Evolution: Dynamical Ordering in Sequence Space." In *Emerging Synthesis in Science, Proceedings of the Founding Workshops of the Santa Fe Institute*, edited by D. Pines. Santa Fe Institute Studies in the Sciences of Complexity, Proc. Vol. I. Reading, MA: Addison-Wesley, 1987, 25–69.

16. Eigen, M. "New Concepts for Dealing with the Evolution of Nucleic Acids." In *Cold Spring Harbor Symposia on Quantitative Biology*, vol. LII. New York, NY: Cold Spring Harbor Laboratory, 1987, 307–320.

17. Eigen, M., and P. Schuster. *The Hypercycle: A Principle of Natural Self-Organization*. New York, NY: Springer-Verlag, 1979.

18. Ewens, W. *Mathematical Population Genetics*. New York, NY: Springer-Verlag, 1979.

19. Fieser, T. M., J. A. Tainer, H. M. Geysen, R. A. Houghten, and R. A. Lerner. "Influence of Protein Flexibility and Peptide Conformation on Reactivity of Monoclonal Anti-Peptide Antibodies with a Protein Alpha-Helix." *Proc. Natl. Acad. Sci. USA* **84** (1987):8568–8572.

20. Fish, S., and T. Manser. "Influence of the Macromolecular Form of a B Cell Epitope on the Expression of Antibody Variable and Constant Region Structure." *J. Expl. Med.* **66** (1987):711–724.

21. Fontana, W., and P. Schuster. "A Computer Model of Evolutionary Optimization." *Biophys. Chem.* **26** (1987):123–147.

22. Getsoff, E.D., H. M. Geysen, S.J. Rodda, H. Alexander, J. A. Tainer, and R. A. Lerner. "Mechanisms of Antibody Binding to a Protein." *Science* **235** (1987):1191–1196.

23. Geysen, H. M., S. J. Barteling, and R. H. Meloen. "Small Peptides Induce Antibodies with a Sequence and Structural Requirement for Binding Antigen Comparable to Antibodies Raised Against the Native Protein." *Proc. Natl. Acad. Sci. USA* **82** (1985):178–182.

24. Geysen, H. M., S. J. Rodda, and T. J. Matson. "The Delineation of Peptides Able to Mimic Assembled Epitopes." *CIBA Foundation Symposium* **119** (1986):130–149.

25. Geysen, H. M., S. J. Rodda, and T. J. Matson. "Strategies for Epitope Analysis Using Peptide Synthesis." *J. Immunol. Methods.* **102** (1987):259–274.

26. Gillespie, J. H. "A Simple Stochastic Gene Substitution Model." *Theor. Pop. Biol.* **23(2)** (1983):202–215.

27. Gillespie, J. H. "Molecular Evolution over the Mutational Landscape." *Evol.* **38(5)** (1984):1116–1129.

28. Heinrich, G., A. Traunecker, and S. Tonegawa. "Somatic Mutation Creates Diversity in the Major Group of Mouse Immunoglobulin K Light Chains." *J. Expl. Med.* **159** (1984):417–435.

29. Honjo, T. "Immunoglobulin Genes." *Ann. Rev. Immunol.* **1** (1983):499–528.
30. Kaartinen, M., G. M. Griffiths, A. F. Markham, and D. Milstein. "mRNA Sequences Define an Unusually Restricted IgG Response to 2-Phenyloxazolone and its Early Diversification." *Nature (London)* **304** (1983):320–324.
31. Karplus, M., and J. N. Kushick. "Dynamics of Proteins: Elements and Function." *Ann. Rev. Biochem.* **53** (1983):263.
32. Karplus, M., A. Y. Brunger, R. Elber, and J. Kuriyan. "Molecular Dynamics: Applications to Proteins." In *Cold Spring Harbor Symposia on Quantitative Biology*, vol. LII. New York, NY: Cold Spring Harbor Laboratory, 1987, 381–390.
33. Kauffman, S. A. "Adaptation on Rugged Fitness Landscapes." In *Lectures in the Sciences of Complexity*, edited by D. Stein, SFI Studies in the Sciences of Complexity, Lect. Vol. I. Redwood City, CA: Addison-Wesley, 1989.
34. Kauffman, S. A. *Origins of Order: Self-Organization and Selection in Evolution.* New York, NY: Oxford University Press, 1990, in press.
35. Kauffman, S. A., and S. Levin. "Towards a General Theory of Adaptive Walks on Rugged Landscapes." *J. Theor. Biol.* **128** (1987):11.
36. Kauffman, S. A., E. D. Weinberger, and A. S. Perelson. "Maturation of Immune Response via Adaptive Walks on Affinity Landscapes." *Theoretcal Immunology, Part I*, edited by A. S. Perelson, Santa Fe Institute Studies in the Sciences of Complexity, Proc. Vol. II. Reading, MA: Addison-Wesley, 1988, 349–382.
37. Macken, C. A., and A. S. Perelson. "Protein Evolution on Rugged Landscapes." *Proc. Natl. Acad. Sci. USA* **86** (1989):6191–6195.
38. Manser, T. L., J. Wysocki, T. Gridley, R. I. Near, and M. L. Gefter. "The Molecular Evolution of the Immune Response." *Immunol. Today* **6** (1985):94–101.
39. Maynard-Smith, J. "Natural Selection and the Concept of a Protein Space." *Nature (London)* **225** (1970):563.
40. Maynard-Smith, J. "The Theory of Games and the Evolution of Animal Conflicts." *J. Theor. Biol.* **47** (1974):209–221.
41. McKean, D., K. Huppi, M. Bell, L. Staudt, W. Gerhard, and M. Weigert. "Gereration of Antibody Diversity in the Immune Response of BALB.c Mice to Influenza Virus Hemagglutinin." *Proc. Natl. Acad. Sci. USA* **81** (1984):3180–3184.
42. Miller, J. H., C. Coulondre, M. Hofer, U. Schmeissner, H. Sommer, and A. Schmitz. "Genetic Studies in the Lac Repressor. IX. Generation of Altered Proteins by the Suppression of Nonsense Mutations." *J. Molec. Biol.* **131** (1979):191–222.
43. Ninio, J. *Approaches Moleculaires de l'Evolution Collection de Biologie Evolution.* New York, NY: Masson, 1979.
44. Perlmutter, R. M. "The Molecular Genetics of Phosphocholine-Binding Antibodies." In *The Biology of Idiotypes*, edited by M. I. Greene, and A. Nisonoff. New York, NY: Plenum Press, 1984, 59–74.

45. Press, J. L., and N. R. Klinman. "Frequency of Hapten Specific B Cells in Neonatal End Adult Mouse Spleen." *Eur. J. Immunol.* **4** (1974):155–159.

46. Roder, et. al. [need reference]

47. Roder, et. al. Personal communication, 1989.

48. Sablitzky, F., G. Wildner, and K. Rajewsky. "Somatic Mutation and Clonal Expansion of B Cells in an Antigen-Driven Immune Response." *EMBO J.* **4** (1985):345–350.

49. Schuster, P. "The Physical Basis of Molecular Evolution." *Chemica Scripta* **26B** (1986):27–41.

50. Schuster, P. "Structure and Dynamics of Replication-Mutation Systems." *Physica Scripta* **35** (1987):402–416.

51. Schlomchik, M.J., D. S. Pisetsky, and M. G. Weigert. "The Structure and Function of Anti-DNA Autoantibodies Derived From a Single Autoimmune Mouse." *Proc. Natl. Acad. Sci. USA* **84** (1987):9150–9154.

52. Shenkin, P. S., D. L. Yarmush, R. M. Fine, H. Wang, and C. Levinthal. "Predicting Antibody Hypervariable Loop Conformation. I. Ensembles of Random Conformations for Ringlike Structures." In *Biopolymers* **26** (1987):55–66.

53. Sherrington, D., and S. Kirkpatrick. "Solvable Model of a Spin Glass." *Phys. Rev. Lett.* **35** (1975):1792.

54. Siskind, G. W., P. Dunn, and J. G. Walker. "Studies on the Control of Antibody Synthesis IrI: The Effect of Antigen Dose and the Supression by Passive Antibody on the Affinity of Antibody Synthesized. *J. Expl. Med.* **127** (1968):55–66.

55. Siskind, G. W., B. Benacerraf. "Cell Selection by Antigen in the Immune Response." *Immunol. Rev.* **10** (1969):1–50.

56. Slaughter, C. A., and J. D. Capra. "Structural and Genetic Basis of the Major Cross-Reactive Idiotype of the a Strain Mouse." In *The Biology of Idiotypes*, edited by M. I. Greene, and A. Nisonoff. New York, NY: Plenum Press, 1984, 75–95.

57. Stein, D. L., G. Baskaran, S. Liang, and M. Barber. "Ground State Structure of Short Range Using Spin Glasses in Two and Three Dimensions." *Phys. Rev. B* **36(10)** (1987):5567–5571.

58. Tonegawa, S. "Somatic Generation of Antibody Diversity." *Nature (London)* **302** (1983):575–581.

59. Wand, A. J. H., H. Roder, and S. W. Englander. "Two-Dimensional HNMR of Cytochrome C: Hydrogen Exchange in the N-Terminal Helix." *Biochemistry* **25** (1986):1107–1114.

60. Weinberger, E. D. "A Rigorous Derivation of Some Properties of Uncorrelated Fitness Landscapes." *J. Theor. Biol.* **134** (1988):125–129.

61. Wright, S. "The Roles of Mutation, Inbreeding, Crossbreeding, and Selection in Evolution." In *Proceedings of the Sixth International Congress on Genetics* **1** (1932):356–366.

62. Wysocki, L., T. Manser, and M. L. Gefter. "Somatic Evolution of Variable Region Structures During an Immune Response." *Proc. Natn. Acad. Sci. USA* **83** (1986):1847–1851.

63. Yancouplos, G. D., and F. W. Alt. "Regulation of the Assembly and Expression of Variable Region Genes." *Ann. Rev. Immunol.* **4** (1986):339–368.

William V. Williams,† Thomas Kieber-Emmons,* David B. Weiner,†‡ and Mark I. Greene‡

†Department of Medicine, University of Pennsylvania School of Medicine, Philadelphia, PA; ‡Department of Pathology, University of Pennsylvania School of Medicine, Philadelphia, PA; and *IDEC Pharmaceuticals Company, La Jolla, CA

Use of Antibodies as Molecular Mimics to Probe Intermolecular Interaction Landscapes

INTRODUCTION

Antibody (Ab)–antigen (Ag) interactions provide well-defined systems for the analysis of intermolecular interactions between large, complex molecules. Strategies for development of an effective immune response have evolved based on genetic principles driven by the need to develop a wide array intermolecular interactions with appropriate specificity and affinity. The genetic machinery utilized by immunoglobulin genes allows tremendous flexibility in selecting antibodies with specific binding properties. These biological properties are manifest in the molecular structure of antibody variable regions.

Each antibody variable region is composed of dimers of a conserved "backbone" structure consisting of anti-parallel beta sheets. From this core structure protrude hypervariable (hv) loops (Figure 1). Three loops are contributed by the light chain, and three by the heavy chain. These hv loops display maximal variability in amino-acid sequence between antibody molecules, and impart the binding specificity for a particular Ag. By analyzing the contribution of these hv loops to the binding interaction individually, the complexity of the intermolecular interactions can be understood at the chemical/non-covalent bond level.

Molecular Evolution on Rugged Landscapes, SFI Studies in the Sciences of Complexity, vol. IX, Eds. A. Perelson and S. Kauffman, Addison-Wesley, 1991

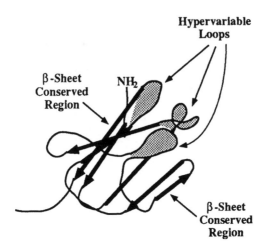

FIGURE 1   Schematic diagram of the general organization of an antibody variable region. The light-chain variable region of an antibody molecule is schematized. The general organization of the region is comprised of an array of anti-parallel beta pleated sheets (backbone region indicated with beta strands shown as heavy arrows, oriented toward the carboxy terminus) which constitute the conserved core. Out of this structure protrude three hypervariable loops, which frequently are reverse turns. In conjunction with these loops other reverse turns protruding from the conserved core have also been implicated as recognition sites.

    The constraints imparted by the conserved portions of the variable region allow development of structural models of individual hv loops. These model structures are useful in analyzing intermolecular interactions implicit in binding. Thus, the complexity of the binding interaction can be reduced so as to appreciate the chemical basis of binding.

    Ab–Ag interactions can be utilized as tools to analyze the binding energetics of a wide range of potential interactions. The ensemble of structures which are recognized by any one Ab can be illustrated as an intermolecular interaction (energy) landscape. This landscape is comprised of molecular species distributed according to their relatedness in structure (which may be the X axis corresponding to "structure space") and chemical properties (the Z axis or "chemical space"). These molecular species, constitute individual "peaks" on the landscape, and the height of each peak is proportional to the binding energy gained upon antibody binding (the Y axis or "energy space") (Figure 2). The molecular species populating this landscape may, however, be different. In molecular terms, the antibody variable region is a large surface with several potential binding sites. Thus, while two distinct molecules may be bound by the same antibody variable region, this does not imply that they both interact with the same sites within this variable region. Thus, the ability of a single monoclonal Ab to bind several Ag may depend on the informational content inherent in the conformational structure of distinct sites within the variable region.

Antibody Variable Region

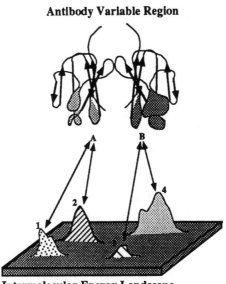

Intermolecular Energy Landscape

FIGURE 2   Schematic of antibody interactions with molecular species on an intermolecular interaction landscape. The landscape consists of structure space (X axis) and chemical space (Z axis) of the potential Ags with the height of the peaks (Y axis or energy space) determined by affinity of binding. The antibody variable region is depicted as having two predominant sites of binding (A and B). These sites are modeled as the predominant sites of interaction with several molecular species (1–4). Thus, the predominant interaction site with molecules 1 and 2 is within variable region site A, while site B is the predominant interaction site with molecules 3 and 4. This representation implies a particular relationship between the universe of molecular species available for recognition and the molecular interaction energy associated with any one molecular structure. This association leads to comparative structural analysis of each of the molecular species that interact with a given antibody, allowing identification of the basis for the relative differences in interaction energetics. This type of approach can lead to the development of new structures with higher affinity which might be of biological interest.

The energetic details of binding are complex, and depend upon many variables. Recognition is dependent on a balance between *intra*molecular interactions lost in the binding process (typified by conformational adjustments of each of the molecular partners and entropy considerations) and energy gain from multiple non-covalent *inter*molecular forces (hydrogen bonding, hydrophobic interactions, aromatic interactions, etc.). In addition an often overlooked component of recognition and binding is the role of solvent. The squeezing out of water from the Ab–Ag interface can contribute greatly to the binding energy. When purely structural/geometric information regarding binding structures are analyzed, it is apparent that several diverse structures can be designed to fit a specific binding region. Antibodies have been developed to mimic other structures, both proteins and non-protein molecules. The utilization of antibodies as molecular mimics allows the development of a constrained system of intermolecular interactions where several points in a binding landscape can be analyzed. The molecular structures involved in binding can be modeled, and the energetics of binding compared.

By developing antibodies to specifically mimic other structures, the diversity of a particular intermolecular interaction landscape can be sampled. Thus, binding in a receptor–ligand interaction can be mimicked by an antibody–receptor interaction; an antibody–ligand interaction; and by an antibody–antibody (idiotype–anti-idiotype) interaction. This is shown schematically in Figure 3. By determining the hv loops involved in binding, and comparing these to the sites involved in the analogous structures, various peaks in a intermolecular interaction landscape can be contrasted. This strategy utilizes applied molecular evolution in selecting antibody variable regions and utilizing them as probes of an intermolecular interaction landscape. Thus, insights are gained into both the particular landscape under investigation and in the strategies utilized by the immune system in evolving interactive molecular structures with defined molecular features. These principles have been applied to the investigation of several ligand receptor interactions. One system, involving receptor binding by a virus, will be dealt with in detail.

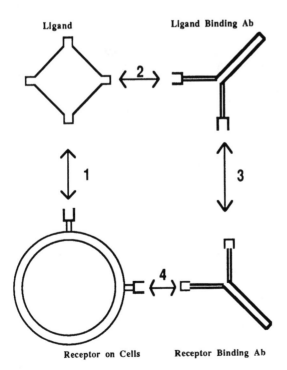

FIGURE 3 Use of antibodies as molecular mimics in ligand–receptor interactions. The ligand receptor binding event (1) is mimicked by three other events (2-4). This involves the use of a ligand binding antibody (which is receptor-like) and a receptor binding antibody (which is ligand-like).

DEVELOPMENT OF AN INTERACTIVE ENERGY LANDSCAPE UTILIZING MIMICRY OF A VIRUS BY AN ANTIBODY

An antibody (87.92.6) was developed to mimic the cell-interaction site of reovirus type 3.[4,5] This was accomplished by immunizing mice with a neutralizing anti-reovirus type 3 monoclonal antibody (9BG5) that mimics the reovirus type 3 receptor (Reo3R, see below), and screening resultant monoclonal antibodies for binding to 9BG5 and Reo3R-bearing cells. 87.92.6 was found to bind specifically to both 9BG5 and murine R1.1 cells (which bear large amounts of Reo3R), as well as to block binding of reovirus type 3 to these cells.[4,5] Thus 87.92.6 represented both an anti-idiotypic antibody and an anti-receptor antibody. It was felt that 87.92.6 contained an "internal image" of the cell-attachment site of reovirus type 3.

Further evidence in support of this was obtained when the amino acid sequence of the 87.92.6 heavy- and light-chain variable regions was deduced from cDNA.[1] These sequences were compared to that of the reovirus type 3 hemagglutinin (HA3), which serves as the cellular attachment protein for reovirus type 3. Sequence similarity was seen between the HA3 and the 87.92.6 variable region. This included similarity between amino acids 323-332 of the HA3 and the second hypervariable region (hvII) of the 87.92.6 light-chain variable region (VL) (Table 1). This raised the possibility that this region of sequence similarity represented the cell-interaction sites of both reovirus type 3 and 87.92.6. This concept was pursued by utilizing synthetic peptides.

By developing synthetic peptides corresponding to the regions of sequence similarity, specific probes were generated for investigations into the significance of the implicated site. A series of experiments were performed with a peptide derived from the 87.92.6 V_L hvII (termed V_L peptide) which encompassed the region of maximal sequence similarity. V_L peptide bound to both 9B.G5 and the Reo3R on cells. This peptide prevented both 87.92.6 and reovirus type 3 from binding.[8] This and

TABLE 1 Amino Acid Sequence Similarity Between 87.92.6 and the Reovirus HA3

Designation	Sequence[1]									
Reovirus HA3	Ile	Val	Ser	Tyr	Ser	Gly	Ser	Gly	Leu	Asn
	+	+	o	o	o	o			o	+
87.92.6 V_L hvII	Leu	Leu	Ile	Tyr	Ser	Gly	Ser	Thr	Leu	Gln

[1] Amino acids that are identical are marked with an (o) while those of the same class are marked with a (+).

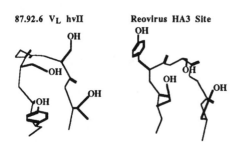

87.92.6 V_L hvII   **Reovirus HA3 Site**

FIGURE 4 The putative binding sites of the 87.92.6 antibody and reovirus type 3 are depicted (artists rendition of molecular models). The hydroxyl groups (OH) in bold print are felt to be involved in hydrogen bond formation needed for binding to the receptor.

other evidence[10,11] suggests that the region of sequence similarity is an important site involved in interaction with the Reo3R and 9B.G5. Molecular models for these sites were developed, and are represented in Figure 4.

As can been seen in Figure 4, the binding site models are similar in the spatial distribution of hydroxyl groups believed to be directly involved in the binding interaction. This is a reflection of the similar amino acids utilized by the virus and antibody in constructing the binding site (see Table 1). When these binding sites are utilized to develop an interactive energy landscape, their structural similarity would place them in close proximity in "sequence space" (i.e., similar amino acids are utilized). Binding studies seem to indicate a higher affinity of binding for the virus compared with the antibody.[11] Thus, the energy of binding for the viral site may be higher than the energy of binding of the antibody site. This is modeled as a higher peak on the energy landscape for the viral site. The sort of landscape developed in this system is depicted in Figure 5.

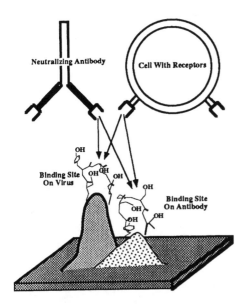

FIGURE 5 Intermolecular interaction landscape based on the molecular features of the viral and antibody binding sites. The proximity of the peaks is due to the structural and chemical similarity between the sites, with the height of the peaks determined by the interaction energies with the receptor and neutralizing antibodies.

We learn several cogent points from the binding strategies utilized by the virus and antibody. In terms of molecular evolution, this is an example of convergence. Both the virus and the antibody employ similar amino acids in constructing their binding sites. They are both arrayed in similar structures (reverse turns) and are likely to bind their receptors in a similar fashion. These observations may stem from the virus utilizing a reverse turn structure for binding. As noted above, antibodies commonly employ reverse turns in their binding regions. Thus, several constraints were imposed on the antibody in constructing a binding region to mimic that of the virus: the same 20 amino acids were available to both virus and antibody; similar geometric structures had to be developed in both cases; and both virus and antibody had to bind the same receptor and the same neutralizing antibody. These constraints are likely to have led to the proximity in sequence space of the two structures. This implies that constraints in structure space may similarly constrain binding mimics in sequence space. Whether this principle applies in a general fashion to other intermolecular interactions requires experimental verification.

DEVELOPMENT OF AN INTERACTIVE ENERGY LANDSCAPE UTILIZING MIMICRY OF A RECEPTOR BY AN ANTIBODY

The reovirus system described above allows development of an additional intermolecular interaction landscape. This landscape is populated by the receptor and the antibody that mimics it (see Figure 3). The 9B.G5 antibody that mimics the receptor for the virus was developed to interact with the receptor binding site on the virus.[2,7] The major constraints imposed on the development of this antibody were that it bound specifically to reovirus type 3, and that it neutralized reovirus type 3 infectivity (i.e., prevented reovirus type 3 from binding the Reo3R). It was later shown that 9B.G5 mimics the Reo3R by several additional criteria. Among these are the ability of 9B.G5 to bind 87.92.6 and other antibodies that mimic reovirus type 3, and the ability of 9B.G5 to bind small peptides that also bind the Reo3R.[8] Thus, the amino acids comprising the variable region of 9B.G5 mimic several interactions displayed by the Reo3R. The structure of the 9B.G5 variable region has not yet been determined, but is likely to be similar to other antibody variable regions in a general sense.

The Reo3R on some cells has been characterized.[3,6] These studies reveal that N-acetylneuraminic acid (sialic acid) is a potential receptor for reovirus type 3. Evidence for this includes that removal of sialic acid from cells decreases reovirus type 3 binding, and that sialic acid can inhibit binding of reovirus type 3 to cells. It is therefore of interest to develop interaction schemes with sialic acid as a model structure. This has been performed by computer modeling, and the cogent features of the interaction scheme are diagrammed in Figure 6.

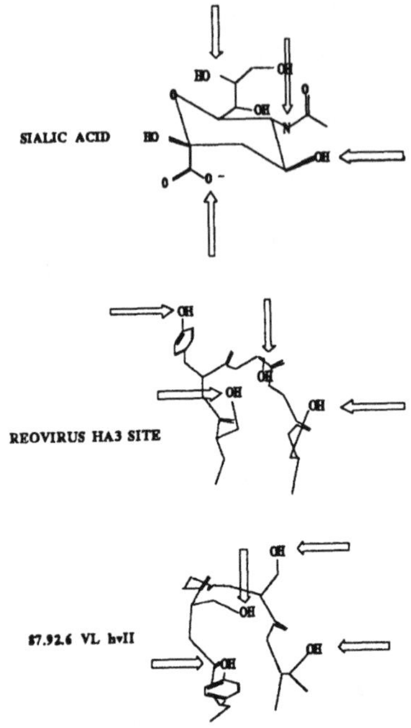

FIGURE 6 Artist's depiction of sialic acid, the reovirus type 3 receptor binding site, and the 87.92.6 V_L hvII. Sialic acid is implicated as s receptor for reovirus type 3. The potential sites of interaction on sialic acid (via hydrogen bonding) with the corresponding sites on the virus and the hvII are indicated by arrows.

This interaction scheme has been probed by utilizing variants of V_L peptide with deletion of potential hydrogen bond sites. Deletion of any of the hydroxyl groups indicated in Figure 6 decreases the affinity of the corresponding peptide for the Reo3R. Thus, all are likely involved in hydrogen bond formation as noted. In contrast, only one of these hydroxyls appears critical for binding to 9B.G5.[9] Thus, while 9B.G5 and sialic acid (as a Reo3R) are capable of interacting with the same small molecular structures, they appear to interact with this site quite differently. This is likely to reflect significant structural differences between 9B.G5 and sialic acid.

Based on these observations, a model landscape can be developed which is populated by sialic acid and the 9B.G5 variable region. In this intermolecular interaction landscape, there is a great distance in structure space between 9B.G5 and sialic acid. Since sialic acid is chemically distinct from any amino acid, the distance in chemical space is also large. The resultant landscape is schematized in Figure 7.

Comparison of this receptor/antibody landscape (Figure 7) with the virus/antibody landscape (Figure 5) has several implications. The additional degrees of freedom gained by expanding the chemical repertoire beyond that of the 20 amino acids may release many constraints in structure and interaction. Specifically, the sialic acid moiety is chemically distinct from the amino acids utilized by 9B.G5, and the structure of sialic acid and the 9B.G5 variable region are quite different. This structural difference is reflected in the manner which these two molecular species bind. Sialic acid utilizes (at least) 4 hydrogen bonds to interact with 87.92.6/reovirus type

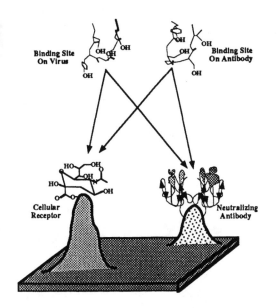

FIGURE 7 Intermolecular interaction landscape populated by sialic acid (as a Reo3R) and the 9B.G5 variable region. The height of the peaks is related to the interaction energies with reovirus type 3 and 87.92.6.

3 while 9B.G5 utilizes only one critical hydrogen bond. These differences imply that while common principles may be involved in the formation of non-covalent bonds, these principles can lead to significantly different binding structures and geometries. However, if structural or chemical constraints are present in such a system, the corresponding intermolecular interactions may develop in a constrained (and to some extent predictable) manner.

CONCLUSIONS

These studies indicate that when molecular evolution takes place in a constrained system, these constraints may be reflected in a relatively convergent process. Such a process will lead to similar structures and binding strategies. However, if molecular evolution is allowed to proceed in a system wherein additional degrees of freedom are present, divergence is likely to predominate. In this case the molecular structures involved in binding and the specific intermolecular interactions may be quite dissimilar. When examining the results of a molecular evolutionary process, the convergence or divergence of the molecular species developed thereby directly reflects the extent of the constraints acting during the evolutionary process. In designing a molecular evolutionary strategy, the constraints imparted on the system will thereby determine the nature of the products that are developed.

ACKNOWLEDGMENTS

William V. Williams is the recipient of an NIH First Award; David B. Weiner is supported by grants from the American Foundation for AIDS research and the Council for Tobacco Research; Mark I. Greene is the recipient of grants from NIH, the National Cancer Institute, the American Cancer Society, National Eye Institute, the Lucille Markey Foundation, and the American Council for Tabacco Research.

REFERENCES

1. Bruck, C., M. S. Co, M. Slaoui, G. N. Gaulton, T. Smith, B .N. Fields, J. I. Mullins, and M. I. Greene. "Nucleic Acid Sequence of an Internal Image-Bearing Monoclonal Anti-Idiotype and Its Comparison to the Sequence of the External Antigen." *Proc. Natl. Acad. Sci. USA* **83** (1986):6578–6582.
2. Burstin, S. J., D. R. Spriggs, and B. N. Fields. "Evidence for Functional Domains on the Reovirus Type 3 Memagglutinin." *Virology* **117** (1982):146–155.
3. Gentsch R. J., and A. F. Pacitti. "Effect of Neuraminidase Treatment of Cells and Effect of Soluble Glycoproteins on Type 3 Reovirus Attachment to Murine L Cells." *J. Virol.* **56** (1985):356–364.
4. Kauffman, R. S., J. H. Noseworthy, J. T. Nepom, R. Finberg, B. N. Fields, and M. I. Greene. "Cell Receptors for Mammalian Reovirus. II. Monoclonal Anti-Idiotypic Antibody Blocks Viral Binding to Cells." *J. Immunol.* **131** (1983):2539–2541.
5. Noseworthy, J. H., B. N. Fields, M. A. Dichter, C. Sobotka, E. Pizer, L. L. Perry, J. T. Nepom, and M. I. Greene. "Cell Receptors for the Mammalian Reovirus. I. Syngeneic Monoclonal Anti-Idiotypic Antibody Identifies a Cell Surface Receptor for Reovirus." *J. Immunol.* **131** (1983):2533–2538.
6. Pacitti A. F., and J. R. Gentsch. "Inhibition of Reovirus Type 3 Binding to Host Cells by Sialylated Glycoproteins is Mediated through the Viral Cell Attachment Protein." *J. Virol.* **61** (1987):1407–1415.
7. Spriggs, D. R., K. Kaye, and B. N. Fields. "Topological Analysis of the Reovirus Type 3 Hemaglutinin." *Virology* **127** (1983):220–224.
8. Williams, W. V., H. R. Guy, D. H. Rubin, F. Robey, J. W. Myers, T. Kieber-Emmons, D. B. Weiner and M. I. Greene. "Sequence of the Cell-Attachment Sites of Reovirus Type 3 and Its Anti-Idiotypic/Antireceptor Antibody: Modeling of Their Three-Dimensional Structures." *Proc. Natl. Acad. Sci. USA* **85** (1988):6488–6492.
9. Williams, W. V., T. Kieber-Emmons, D. B. Weiner, and M. I. Greene. "Contact Residues and Predicted Structure of the Reovirus Type 3-Receptor Interaction." Presented at Modern Approaches to New Vaccines Including the Prevention of AIDS, September 14-18, Abstract No. 123, Cold Spring Harbor Labs, NY, 1988.
10. Williams, W. V., S. L. London, D. H. Rubin, S. Wadsworth, D. B. Weiner, J. A. Berzofsky, and M. I. Greene. "Immune Response to a Molecularly Defined Internal Image Idiotope." *J. Immunol.* **142** (1989):4392–4400.
11. Williams, W. V., D. A. Moss, T. Kieber-Emmons, J. A. Cohen, J. N. Myers, D. B. Weiner, and M. I. Greene. "Development of Biologically Active Peptides Based on Antibody Structure." *Proceedings of the National Academy of Sciences USA* **86** (1989):5537–5541.

Gérard Weisbuch† and Alan S. Perelson‡
†Laboratoire de Physique Statistique de l'Ecole Normale Supérieure, 24 rue Lhomond, F 75231, Paris Cedex 5, France and ‡Theoretical Division, Los Alamos National Laboratory, Los Alamos, NM 87545

Affinity Maturation and Learning in Immune Networks

1. INTRODUCTION

The metaphors of cognition are often used when discussing the properties of the immune system: one speaks of recognition of antigen or of the memory of a previous encounter with antigen. Here we will compare some "cognitive" properties of immune network models with those of neural network models.[10] Both sets of models are based on the analysis of the dynamical properties of networks of automata or of systems of differential equations.[1,7,16,20,21] Attractors of the dynamics are interpreted in terms of memories that can be retrieved by inputting some "pattern" to a neural network or giving antigen to an immune network. The appearance of new attractors is interpreted as learning. New attractors can appear for two reasons:

- If the structure and parameters of the network are kept fixed, new attractors appear when the system is driven into regions of the configuration space that had not been previously visited. This is described in most models of immune memory when encounter with antigen results in reaching previously existing

Molecular Evolution on Rugged Landscapes, SFI Studies in the Sciences of Complexity, vol. IX, Eds. A. Perelson and S. Kauffman, Addison-Wesley, 1991   **189** attractors.[1,8,20] Hoffman[9] developed a neural network model based on an analogy with the immune system that also exhibits this type of learning.

- In most neural network models the parameters describing the network are changed during learning. Thus, the degree of interaction or synaptic weights between neurons are modified by pattern presentations: new potential attractors appear that can be reached by the network dynamics.[10]

The purpose of this paper is to show that in the immune system affinity maturation via somatic mutation can be interpreted in terms of a learning mechanism of the second kind. We will first very briefly recall the formalism used to describe learning in simple neural networks, the so-called Hebb's rule. We will then describe qualitatively the possible existence of a similar rule for the immune system. This will be further checked with a simple model. Finally, learning in neural networks and immune networks will be compared.

LEARNING IN NEURAL NETWORKS

In their simplest version,[10] neural networks are composed of model neurons, threshold automata, which update their state, S_i, at each time-step according to a threshold function:

$$S_i = Y\left(\sum_{j=1}^{N} J_{ij} S_j - \Theta_i\right), \qquad i = 1, 2, \ldots, N, \qquad (1)$$

where Y is a Heavyside function, equal to 1 if its argument is positive and 0 otherwise. The quantity $\sum_j J_{ij} S_j$ is sometimes called the local field h_i acting upon automaton i. Automata j connected to automaton i contribute to the field with different connection intensities J_{ij}. If the field is above the threshold Θ_i, $S_i = 1$; otherwise $S_i = 0$. If the J_{ij} are symmetrical, i.e., $J_{ij} = J_{ji}$, and the system is updated randomly, i.e., one randomly chosen automaton updates its state at each time step, then the system will ultimately settle down to a fixed state, called a fixed point. The final state that the system reaches is called an attractor. In this system all of the attractors are fixed points. There may be more than one attractor, and which one the system ultimately reaches will depend on the initial condition and the choice of J_{ij}.

A learning algorithm which updates the J_{ij} allows the network to "memorize" some chosen configurations as attractors of the dynamics. A list (or vector) of the states for each neuron in the network is called a pattern. When a pattern to be learned, $S^p = (S_1^p, S_2^p, \ldots, S_N^p)$, is presented to the network, the connection strengths are changed according to the rule

$$J_{ij} \longrightarrow J_{ij} + (2S_i^p - 1)(2S_j^p - 1). \qquad (2)$$

This prescription is called Hebb's rule. During the learning process, when two connected automata are in the same state (either one or zero) their connection intensity is increased. It is decreased when they are in a different state. In the retrieval phase, normal dynamics with fixed J_{ij} are performed; presentation of a pattern, somewhat different from the reference pattern S^p results in driving the network towards S^p or an attractor very similar to S^p.

AFFINITY MATURATION AS LEARNING
QUALITATIVE ANALYSIS

A number of models in immunology[1,4,5,20,21] relate the efficiency of antigen or anti-idiotype recognition by clone i to a field h_i that is a linear combination of the concentrations of antigen and anti-idiotypic clones j recognized by clone i, weighted by the affinity of each interaction. To be precise, the field is given by the following expression

$$h_i = \sum_j J_{ij} x_j + J_{ia}\, a \quad , \tag{3}$$

where the J_{ij}'s are scaled affinity constants representing the influence of clones j with population size x_j, on clone i. The J_{ij}'s are usually taken to be symmetric: $J_{ij} = J_{ji}$. The second term is built on the same principle and represents the influence of antigen with concentration a.

For a given set of clones the J_{ij}'s are constant. However, during the course of an immune response, antibody variable region genes within stimulated B cells mutate rapidly so that a single clone generates a diverse set of subclones. Instead of considering the J_{ij}'s of individual clones, let us clump together as a quasi-species the sub-clones that have been generated by somatic mutation from a given germline gene. (Here we use the term quasi-species in the same sense as Eigen and Schuster.[6]) The simplest way to define the affinity constant of a quasi-species would be to average the J_{ij}'s over the germ-line and mutant populations:

$$\langle J_{ij} \rangle = \frac{\sum_{i'} J_{i'j} x_{i'}}{\sum_{i'} x_{i'}} \quad , \tag{4}$$

where i' refers to those sub-clones with population size $x_{i'}$ that derive from the germ-line by somatic mutations. The two summations also include the germ-line clone with population size x_i. In the course of affinity maturation, mutants with higher affinity constants are selected, presumably because they continue to proliferate even as the antigen concentration decreases (cf. Siskind and Benacerraf[17]). Thus populations with high affinity increase in size more than that of mutants with lesser affinities. In other words, when antigen is presented, the average J_{ij} will increase; this reminds us of Hebb's rule.

MATHEMATICAL ANALYSIS

We describe the immune response by a simple ordinary differential equation model derived from previous works of Gunther and Hoffmann,[7] Segel and Perelson,[16] De Boer and Hogeweg,[1] and Weisbuch,[20] among others. To keep the model simple we do not make any distinction between antibody and B-cell responses. We thus consider elements of only one type, say, B lymphocytes. We imply that antibody concentrations follow directly from the cell populations that are the variables of the model. See Perelson,[14] Segel and Perelson,[16] Stewart and Varela,[18] De Boer and Hogeweg[2,3] and De Boer et al.[5] for models that deal separately with antibodies and B cells. The cells in the model are characterized by their idiotype, and the basis of the model is the following system of differential equations describing the evolution of the different germ-line cell populations x_i and mutant populations $x_{i'}$

$$\frac{dx_i}{dt} = s + x_i(rf(h_i) - d - m) + \sum_{i'} m_{ii'}(t)x_{i'} \qquad (5a)$$

$$\frac{dx_{i'}}{dt} = x_{i'}(rf(h_{i'}) - d - m) + m_{i'i}(t)x_i \qquad (5b)$$

The first term in Eq. (5a) is a source term corresponding to incoming cells from the bone marrow. The second term is the net proliferation rate. It incorporates three parts. The first, $rf(h_i)$, is the specific rate of growth of activated cells, r, multiplied by the fraction of cells activated to grow, $f(h_i)$. The second part, $-d$, is the natural death rate, and the third part, $-m$, is the rate of loss of cells in population i due to somatic mutation. If we let $m_{i'i}$ be the rate at which a cell of type i mutates into a cell of type i', then $m = \sum_{i'} m_{i'i}$. During affinity maturation, the mutations that are observed are point mutations. Thus, back mutations should be possible, especially among one-mutant neighbors, i.e., sequences that differ from one another by a single point mutation.[12] The last term in Eq. (5a) accounts for back mutation, where $m_{ii'}$, is rate of creation of cells of type i by mutation of clones i'. Equation (5b) has a structure similar to that of Eq. (5a), except no bone marrow source term, s, is present, and we have assumed that only the mutation of the germ-line clone can give rise to the mutant i' clone. This would be the case, for example, if all of the somatic mutants were one-mutant neighbors of the germ-line clone.

Several mathematical expressions have been proposed for the function $f(h_i)$ describing the fraction of cells of type i activated to grow. We will assume that this function is the same for all cells, i.e., $f(h_i) = f(h_{i'})$ for all i'. In most cases f is a bell-shaped function of h_i—reflecting, for example, measured stimulus response curves as a function of antigen concentration. Receptor cross-linking, which is important in B-cell stimulation, also gives rise to bell-shaped activation functions (cf. Perelson and DeLisi[15]; Perelson[13]). The important point is that the response is weak when

h_i is small, large in the intermediate region, and small again when h_i gets large. $f(h)$, whose graph is shown in Figure 1, is defined by:

$$f(h) = \left[\frac{h}{h+\theta_1}\right]\left[1 - \frac{h}{h+\theta_2}\right] = \frac{h\theta_2}{(h+\theta_1)(h+\theta_2)} \quad . \tag{6}$$

We now summarize previous results[1,20,21] on the dynamics of the system in the absence of mutations. If $x_{i'}(0) = 0$ and $m_{i'i} = 0$, then only germ-line clones exist and Eq. (5) reduces to

$$\frac{dx_i}{dt} = s + x_i(rf(h_i) - d) , \tag{7}$$

The basic principle of anti-idiotypic control of the immune response to a given antigen is simply described by considering a pair of clones, X specific for the antigen,

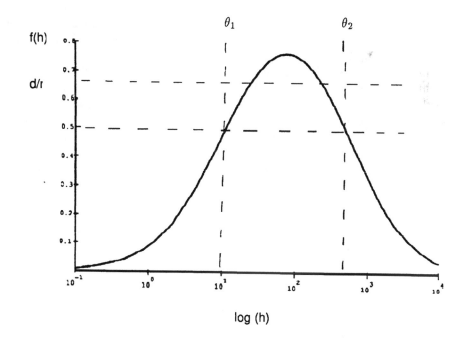

FIGURE 1   Graph of $f(h)$, the activation function for cell growth. h is the field acting upon idiotype i due to other idiotypes and antigen, and θ_1 and θ_2 are the proliferation and suppression thresholds. This curve was generated with $\theta_1 = 10$ and $\theta_2 = 500$.

and Y anti-idiotypic to X. Equation (7) then reduces to a set of two equations for the population sizes of X and Y, which we call x and y, respectively. The only interaction constant is $J = J_{12} = J_{21}$. The field of X, $h_1 = Jy$ and the field of Y, $h_2 = Jx$. Antigen encounter is modeled by an increase of the X population to some high level. This is then chosen as the initial condition for the model; as in the rest of this text, we do not model the dynamics of the antigen removal. The system has three stable steady states:

1. Prior to antigen encounter one expects the system to be in a "virgin state". In this state both clones remain unactivated, i.e. $f(h_i) \simeq 0$, and consequently the populations sizes of clones X and Y are given by:

$$x_0 = y_0 \simeq \frac{s}{d} \ , \tag{8}$$

where the subscript 0 is used to denote the virgin state. This equation is only an approximation because f is not precisely zero. This state is a stable attractor.

2. After the system has been presented with antigen specific to clone X, the X population rises to a level that allows it to escape from the basin of attraction of the virgin state and a new attractor is reached with a "high" population of X and a "low" population of Y. The population sizes of X and Y in this state are both higher than in the virgin state. This state is maintained even when the antigen concentration is depleted, and thus we call it a "vaccinated" or "immune" state. In this state, the idiotypic clone X is maintained at a high population level by stimulation of the anti-idiotypic clone Y. The field that influences clone X, as shown in Figure 1, corresponds to the rising part of the $f(h)$ curve, and the steady-state level of Y is determined approximately by the first intersection of $f(h)$ with d/r (see below). The anti-idiotypic clone Y is maintained at a low population level. It is being overstimulated or suppressed by the high level of clone X. Its field is on the declining part of the $f(h)$ curve, and the steady-state level of X is then determined approximately by the second intersection of $f(h)$ with d/r.

In the limit, when the number of cells provided by the bone marrow source, s, is sufficiently small compared with the number of cells provided by proliferation, the following simple expression results for the steady-state population levels of clones X and Y in the immune state:

$$x_1 = \frac{\theta_2}{J} \frac{r-d}{d} \ , \tag{9a}$$

$$y_1 = \frac{\theta_1}{J} \frac{d}{r-d} \ . \tag{9b}$$

This expression is obtained by equating the proliferation rate $rf(h_i)$ with the death rate d, and by approximating $f(h_1)$ by $h_1/(h_1+\theta_1)$ and $f(h_2)$ by $\theta_2/(\theta_2+h_2)$, according to whether one is interested in either the first or the second intersection of the curve $f(h)$ with the line $d/r = $ constant. From Eq. (6) one can see that $f(h)$ can be approximated by the first term in brackets if

$h \ll \theta_2$ and by the second term in brackets if $h \gg \theta_1$. In the steady state under consideration, x is high and y low. Thus $h_1 = Jy$ is low and the first intersection is the relevant one for computing $f(h_1)$. Analogously, $h_2 = Jx$ is high and the second intersection is the relevant one for computing $f(h_2)$.

3. The third attractor is an "anti-immune state." Because the equations are symmetric in x, y, it is found by exchange of X with Y.

We now take into account the influence of somatic mutation. and analyze the model given by Eq. (5). The idiotypic clone X is now replaced by a quasi-species—a set composed of the germ-line clone plus a number of closely related mutants. The quasi-species of the anti-idiotype Y also contains a germ-line clone and the same number of mutants. To illustrate, we have performed numerical simulations with two somatic mutants of X and Y. Let X_1 and Y_1 denote the two germ-line clones (X and Y), and let X_2 (Y_2) and X_3 (Y_3) denote their somatic mutants. For the J_{ij}'s, we somewhat arbitrarily suppose that the affinity relating a clone X_i and its anticlone Y_j is given by:

$$J_{ij} = .5(i + j) \ , \tag{10}$$

where i and j vary from 1 to 3. Since the two germ-line populations were numbered 1, their mutual affinity is the lowest possible in the system, i.e., $J_{11} = 1$. In this scheme, the mutant clones, numbered 2 and 3, have higher affinities and they range from 1.5 (J_{12}) to 3 ($J_{3,3}$).

Numerical solution of Eqs. (5) show that the system evolves toward an attractor where the idiotype with the highest affinity and anti-idiotype with the lowest affinity is selected. A typical solution is represented in Figure 2.

Simple reasoning allows one to predict approximate values of the sub-clone populations at the attractor. In the limit of small bone marrow and mutational source terms, a steady state is reached when $rf(h)$ is close to d, i.e., when birth by proliferation matches death, as is the case in the model given by Eq. (7) for two germ-line populations. When a higher affinity somatic mutant appears, its field is initially higher than that of the germ-line clone since the terms in h, Eq. (3), are products of affinities and population sizes. This larger field induces a larger value of $rf(h)$, the growth rate, and the mutant population starts increasing. Analogously, the populations of anti-idiotypes, which are being suppressed by the idiotypic clones and operate near the second intersection of $f(h)$ with d/r, decrease because they experience a higher field (see Figure 1). This decrease limits the growth of the idiotypic populations. Equilibrium is achieved when the highest proliferation functions stabilize at d/r. In the low field region, for the idiotypes, the clone with the largest J_{ij} has the highest proliferation function and preferentially grows. In the high, suppressive, field region for the anti-idiotypes, the highest proliferation function corresponds to the lowest field, and thus the clone with the lowest J_{ij} grows best. The selection of specific populations is quite strong, as can be checked by using the iterative replacement approach described below.

For any clone, steady state is given by

$$x_i = \frac{s_i}{d' - rf(h_i)} \ , \tag{11}$$

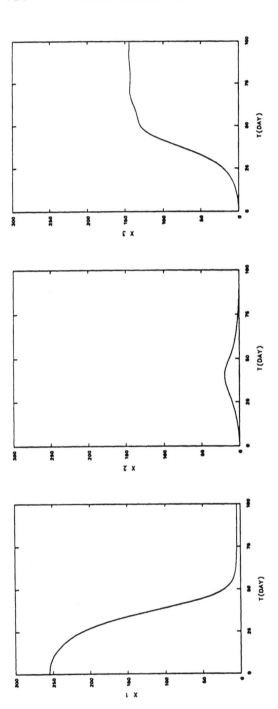

FIGURE 2 Maturation of the immune response under the influence of somatic mutation. Time evolution of germ-line idiotype (X_1), mutant idiotypes (X_2 and X_3) and gram-line anti-idiotype (Y_1) and mutant anti-idiotype (Y_2 and Y_3) populations. Notice each graph is plotted on a different scale. The germ-line idiotype population, first excited by antigen to a high level at $t = 0$, is replaced by its highest affinity somatic mutant, X_3. The anti-idiotype germ-line clone (Y_1) maintains its predominance among the set of anti-idiotype mutants because it has the lowest affinity. Its population decreases because its affinity for the dominant idiotype "species" has increased. Simulation parameters were $\theta_1 = 10$, $\theta_2 = 500$, $s = 1\ \mathrm{d}^{-1}$, $r = 1.5\ \mathrm{d}^{-1}$, $d = 1\ \mathrm{d}^{-1}$, $m = 0.001\ \mathrm{d}^{-1}$, and $m_{12} = m_{13} = 0.0005\ \mathrm{d}^{-1}$. The interaction constants J_{ij} obey Eq. (10). The simulation was started with the system in the immune state and only the two germ-line clones had non-zero population sizes.

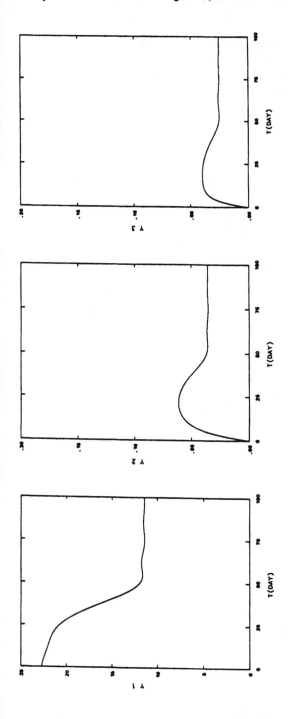

FIGURE 2   continued

where $d' = d+m$, and s_i is a generalized source term equal to the positive mutation term in Eq. (5) plus, for germ-line clones, their intrinsic rate of production from the bone marrow, s. For those sub-clones that have affinities smaller than the highest, the denominator in Eq. (11) can be evaluated with reference to f_0, the highest value of $f(h)$:

$$f(h) = f_0 - \Delta f \ ,$$

and the expression of x_i rewritten:

$$x_i = \frac{s_i}{d' - rf_0} \frac{d' - rf_0}{d' - r(f_0 + \Delta f)} \ . \tag{12}$$

The first fraction is just the steady-state population size of the fittest sub-clone of the quasi-species. The second fraction is usually much smaller than 1, since f_0 is much closer to d' than any other f. In the immune state, the population sizes of the fittest sub-clones can be computed as in the two-clone model. Thus, if the source of cells is small, these population sizes are still given by Eq. (9), where J is the highest interaction constant, and d is replaced by d'. The steady-state population sizes of clones with lower interaction constants are computed from Eq. (11). To compute x_i, $i = 1$, 2 for instance, the denominator of Eq. (11) is approximated by expanding $f(h)$ in the vicinity of f_0 to first order:

$$d' - rf(h_i) = -r\Delta f = -rf'_h \Delta h = \frac{r\theta_1}{(h_i + \theta_1)^2} \Delta J y_1 \ , \tag{13}$$

where f'_h is the derivative of f with respect to h evaluated at the field of f_0, and ΔJ is the deviation from the maximum interaction constant J. Substituting Eq. (9b) for y_1 and recognizing that $\theta_1/(h_i + \theta_1) = 1 - f(h_i) = 1 - d'/r$, one finds

$$x_i = \frac{s_i J}{d' \Delta J} \frac{r}{r - d'} \ . \tag{14}$$

The same method shows that for anti-idiotypic clones with interaction constants lower than the maximum, $y_i = x_i$. These approximate expressions are in reasonable agreement with our simulation results (Figure 2).

These preliminary results show that because of somatic mutation:

- High-affinity idiotypic sub-clones are strongly selected from a population containing clones with a lesser affinity. As a result, the affinity of the quasi-species increases. This increase in affinity occurs along with a "virtual" decrease of the quasi-species' population size in the resulting immune steady state with respect to the population size that would be established in the absence of hypermutation. (In Figure 2, x_3 does not rise to the original x_1 level.) This latter effect would only be observable if the hypermutation process were much slower that the build-up of the immune response.

- Low-affinity anti-idiotypic sub-clones are selected because they receive less suppression. The same population depression as for idiotypic clones is observed (see y, curve in Figure 2).

Our results should only be considered preliminary due to our particular choice of the J_{ij}'s [Eq. (10)]. But we still expect the previous algebraic analysis to remain valid in the case of a more complex J_{ij} "landscape."

CONCLUSIONS

Comparing Hebb's rule, or other versions of learning algorithms in neural networks, to affinity maturation results in the following similarity and differences.

The main similarity is the change of interaction constants due to pattern presentation or antigen encounter. But learning by increase of J_{ij} in the immune system only occurs when both populations i and j are large. This is in contrast to Hebb's rule in which learning also occurs when both connected automata are in the zero state. Thus learning in the immune system seems to correspond to a "modified Hebb's rule":

$$J_{ij} \longrightarrow J_{ij} + \epsilon x_i x_j \ , \tag{15}$$

where ϵ is some learning constant.

Of course the basic processes of learning in both systems are different. In the nervous system, learning is most probably due to physico-chemical changes occurring at the level of synapses, while in the immune system, "learning" is due to a selection process among novel clones generated by somatic mutation.

The presentation of a pattern to a neural net implies the excitation of a large fraction of the elements of the networks. This is not the case in the immune system where antigen should only perturb the few relevant clones that recognize it, plus their anti-idiotypic partners. In some cases, which we believe to be pathological, the anti-idiotypic clones could excite anti-anti-idiotypic clones, etc. leading to "percolation" and generalized activation of the immune system.[2,14] In general, we believe that the immune response is localized in the space of idiotypes.[20]

We have not investigated the kinetics of learning by affinity maturation as compared with the kinetics of the immune response. In neural networks it is generally assumed that the pattern recognition process (dynamics of the S variables with constant J_{ij}'s) is faster than the learning process. This assumption might not be true for immune networks.

NOTE IN PROOF

Between the time this paper was first sent to the editors and the time the book went to press, further simulations were done with a variety of different J_{ij}'s. The behavior described in this paper is not as general as we had expected. In fact, according to the choice of the J_{ij}'s one can obtain different behaviors corresponding either to the selection of high-affinity idiotypic clones and low-affinity anti-idiotypic clones as described above, or to the coexistence of different mutant clones.

In the limit of small source and mutation terms, the high steady-state populations characteristic of the immune or suppressed states are obtained when the

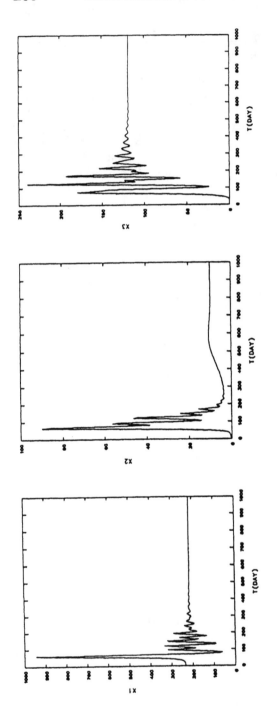

FIGURE 3   Maturation of the immune response under the influence of somatic mutation. The only change in conditions from Figure 2 is the choice of J_{ij}, which is now given by Eq. (19).

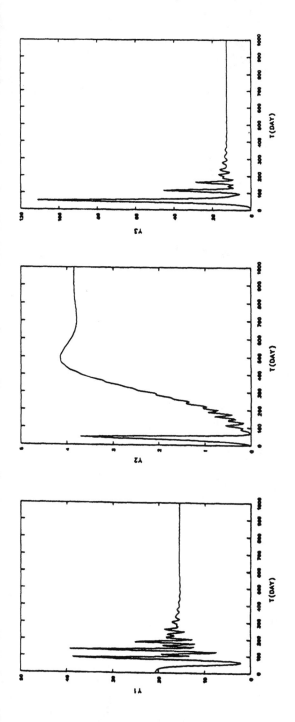

FIGURE 3 continued

proliferation function matches the death rate. Thus the field experienced by each clone obeys an equation such as:

$$h_i = \sum_j J_{ij} x_j = f^{-1}(d'/r) = \text{constant} , \qquad (16)$$

where $f^{-1}(d'/r)$ can be either the low-field or the high-field intersection of $f(h)$ with d'/r.

When J_{ij} is a monotonic function in both i and j, such as $.5(i + j)$, the linear equations (16) in x_j cannot have all their solutions positive. For the case studied in the paper (Figure 2) this system of equations for the fields of the clones X_1, X_2 and X_3 becomes

$$h_{X_1} = y_1 + 1.5y_2 + 2y_3 = 23.1 , \qquad (17a)$$

$$h_{X_2} = 1.5y_1 + 2y_2 + 2.5y_3 = 23.1 , \qquad (17b)$$

$$h_{X_3} = 2y_1 + 2.5y_2 + 3y_3 = 23.1 . \qquad (17c)$$

For clones Y_1, Y_2 and Y_3

$$h_{Y_1} = x_1 + 1.5x_2 + 2x_3 = 216.1 , \qquad (18a)$$

$$h_{Y_2} = 1.5x_1 + 2x_2 + 2.5x_3 = 216.1 , \qquad (18b)$$

$$h_{Y_3} = 2x_1 + 2.5x_2 + 3x_3 = 216.1 , \qquad (18c)$$

where for the parameters of Figure 2, $f^{-1}(d'/r) = 23.1$ or 216.1 depending upon which intersection is being examined. Clearly, with $y_i \geq 0$ and $x_i \geq 0$, only one equation of the set (17) and one of (18) can be satisfied. This implies that two of the X clones and two of the Y clones die out so that their field equations are no longer appropriate to characterize the steady state. As seen in Figure 2 the appropriate solutions are $y_2 = y_3 = x_1 = x_2 = 0$, $x_3 > 0$ and $y_1 > 0$. From the field of X_3, i.e. Eq. (17c) one predicts that at steady state, $y_1 = 11.55$, which is quite close to the actual steady-state value of 11.27. The prediction of the steady-state value of x_3 from Eq. (18a) is not as accurate, but becomes quite accurate for smaller values of the source, i.e., $s < 0.1$.

The affinities, J_{ij}, used above, Eq. (10), can be interpreted in terms of sticky receptors. For example, the receptors can be classified according to their stickiness with respect to all the anti-receptors:

$$J_{i'j} > J_{ij} \quad \text{for all } j \quad \text{if} \quad i' > i .$$

Our simulations show that the stickiest clone is then selected for the idiotype and the least sticky for the anti-idiotype.

Let us now consider another situation, one in which clone X_i interacts most strongly with clone Y_j when $j = i$. Such a situation could arise in a one-dimensional shape-space model spanned by populations of clones occupying complementary positions, such as in the model of Segel and Perelson.[16] In this situation, for instance,

$$J_{ij} = \frac{1 + 0.1i}{1 + (i - j)^2}.$$

(19)

With this choice of J_{ij}, Eq. (16) can be satisfied simultaneously by several clones, idiotypic and anti-idiotypic. Instead of a selection of the fittest, coexistence of several clones is observed along with an oscillatory approach to the steady state (Figure 3). It simply happens that no single clone is fitter with respect to the different anti-idiotypic clones that are present. This diversity corresponds to the existence of several niches.

In conclusion, somatic mutation can result in either selection or coexistence of clones. The present studies are only preliminary. Biological information about the mutual affinities of idiotypic and anti-idiotypic antibodies and their somatic mutants, as well as information about the serum concentrations of these antibodies would be valuable in deciding which behavior is actually occurring. One might also think that the two behaviors described above, plus intermediates in which some clones died out but others coexist, might occur depending upon the antigen and the primed clones that respond to it. A more refined analysis, taking into consideration both stochastic elements and the fact that clone sizes should be integers, might be needed if clones, such as Y_2 in Figure 3, reach sizes close to zero. We are encouraged in this work by the recent findings (Van der Heijden, et al.[19]) of extensive somatic mutations in a human anti-idiotypic antibody following rabies vaccination.

ACKNOWLEDGMENTS

This work was supported by the U.S. Department of Energy, the Santa Fe Institute, NIH Grant AI28433, and by Inserm grant 879002. The Laboratoire de Physique Statistique de l'ENS is associated with CNRS (URA 1306). G. Weisbuch would also like to thank the Santa Fe Institute for their hospitality during his visits and for providing a stimulating intellectual environment for work in Theoretical Immunology.

REFERENCES

1. De Boer, R. J., and P. Hogeweg. "Memory but no Suppression in Low-Dimensional Symmetric Idiotypic Networks." *Bull. Math. Biol.* **51** (1989):223–246.

2. De Boer, R. J., and P. Hogeweg. "Unreasonable Implications of Reasonable Idiotypic Network Assumptions. *Bull. Math. Biol.* 51 (1989):381–408.

3. De Boer, R. J., and P. Hogeweg. "Idiotypic Network Models Incorporating T-B Cell Cooperation: The Conditions for Percolation." *J. Theor. Biol.* 139 (1989):17–38.

4. De Boer, R. J., and A. S Perelson. "Size and Connectivity as Emergent Properties of a Developing Immune Network." *J. Theoret. Biol.*, (1990), in press.

5. De Boer, R. J., I. G. Kevrekidis, and A. S. Perelson. "A Simple Idiotypic Network Model with Complex Dynamics." *Chem. Eng. Sci.*, (1990), in press.

6. Eigen, M., and P. Schuster. *The Hypercycle: A Principle of Natural Self-Organization.* New York: Springer-Verlag., 1979.

7. Gunther, N., and G. W. Hoffmann. "Qualitative Dynamics of a Network Model of Regulation of the Immune System: A Rationale for the IgM to IgG Switch." *J. Theoret. Biol.* 94 (1982):815–855.

8. Hoffmann, G. W. "A Mathematical Model of the Stable States of a Network Theory of Self-Regulation." In *Systems Theory in Immunology*, edited by C. Bruni, G. Doria, G. Koch, and R. Strom. *Lecture Notes in Biomathematics*, vol. 32. Berlin: Springer-Verlag, (1979):239–257.

9. Hoffmann, G. W., M. W. Benson, G. M. Bree, G. M., and P. E. Kinahan. "A Teachable Neural Network Based on an Unorthodox Neuron." *Physica* 22D (1986):233–246.

10. Hopfield, J. "Neural Networks and Physical Systems with Emergent Collective Computational Properties." *Proc. Natl. Acad. Sci. USA* 79 (1982):2554–2558.

11. Jerne, N. K. "Toward a Network Theory of the Immune System." *Ann. Immunol. (Inst. Pasteur)* 125C (1974):373–389.

12. Macken, C. A., and A. S. Perelson. "Protein Evolution on Rugged Landscapes." *Proc. Natl. Acad. Sci. USA* 86 (1989):6191–6195.

13. Perelson, A. S. "Some Mathematical Models of Receptor Clustering by Multivalent Ligands." In *Cell Surface Dynamics: Concepts and Models*, edited by A. S. Perelson, C. DeLisi and F. W. Wiegel. New York: Marcel Dekker, (1984):223–276.

14. Perelson, A. S. "Immune Network Theory." *Immunol. Rev.* 110 (1989):5–36.

15. Perelson, A. S., and C. DeLisi. "Receptor Clustering on a Cell Surface. I. Theory of Receptor Cross-Linking by Ligands Bearing Two Chemically Identical Functional Groups." *Math. Biosciences* 48 (1980):71–110.

16. Segel, L. A., and A. S. Perelson. "Computations in Shape Space. A New Approach to Immune Network Theory." In *Theoretical Immunology, Part Two*, edited by A. S. Perelson. SFI Studies in the Sciences of Complexity, vol. III. Redwood City, CA: Addison-Wesley (1988):321–343.

17. Siskind, G. W., and Benacerraf. "Cell Selection by Antigen in the Immune Response." *Immunol. Rev.* 10 (1969):1–50.

18. Stewart, J., and F. J. Varela. "Exploring the Meaning of Connectivity in the Immune Network." *Immunol. Rev.* 110 (1989):37–61.

19. Van der Heijden, R. W. J., H. Bunschoten, V. Pascual, F. G. C. M. Uyt-dehaag, A. D. M. E. Osterhaus, and J. D. Capra. "Nucleotide Sequence of a Human Monoclonal Anti-Idiotypic Antibody Specific for a Rabies Virus-Neutralizing Monoclonal Idiotypic Antibody Reveals Extensive Somatic Variability Suggestive of an Antigen-Driven Immune Response." *J. Immunol.* **144** (1990):2835–2839.
20. Weisbuch, G. A "A Shape Space Approach to the Dynamics of the Immune System." *J. Theor. Biol.* **143** (1990):507–522
21. Weisbuch, G., R. De Boer, and A. S. Perelson. "Localized Memories in Idiotypic Networks." *J. Theoret. Biol.* (1990), in press.

Gregory W. Siskind
Department of Medicine, The New York Hospital–Cornell Medical Center, 525 East 68th
Street, New York, NY 10021

Auto-Anti-Idiotype and Antigen as Selective Factors Driving B-Cell Affinity Maturation in the Immune System

For many years we have studied the regulation of antibody affinity during the immune response. Measurements of affinity and affinity distributions were carried out both immunochemically by the traditional techniques of equilibrium dialysis,[3] fluorescence quenching,[4] and solid-phase radioimmunoassay,[15] as well as at the antibody-secreting cell level by hapten inhibition of plaque formation.[7] The results obtained by all procedures were biologically comparable and generally consistent with the theory that antigen preferentially stimulates ("selects") high-affinity antibody-secreting cells to proliferate and differentiate into antibody-secreting cells thereby leading to the greater expansion of high-affinity clones relative to lower-affinity clones.[20] This results in a shift in the distribution of B cells, from a population consisting mainly of low-affinity antibody-secreting cells shortly after antigen exposure, to a highly heterogeneous population with a marked representation of high-affinity clones and a high average affinity, late after immunization.[4,6,20,23,24] It should be noted that low-affinity antibodies continue to be produced and their production can be boosted by injection of relatively large doses of antigen.[14,23] In addition to the progressive rise in average affinity.[4,6,23,24] and shift in affinity distribution[23,24] with time after immunization, it was shown that: (1) the rate of increase in affinity is greater after lower doses of antigen[4,6,23,24]; (2) B-cell tolerance preferentially affects high-affinity antibody-producing cells, leading to a reduced average affinity of the antibody produced by partially tolerant animals[21];

(3) boosting with low doses of antigen primarily stimulates a very-high-affinity secondary response; and (4) depletion of T cells leads to reduced affinity maturation in response to T-dependent antigens.[2]

Recent studies in collaboration with Dr. Tova Francus have examined in more detail the role of T cells in the selection of higher-affinity antibody-secreting cells. Cell transfer experiments were used to determine the relative contributions of hapten-primed B cells and carrier-primed T cells to various characteristics of the secondary response: namely, increased affinity and increased IgG antibody with an increased IgG/IgM plaque-forming cell ratio. The shift in isotype expression appears to depend upon a primed B-cell population. In contrast, primed T cells appear capable of rapidly selecting high-affinity B cells following boosting regardless of whether a primed or naive B-cell population is present. This suggests a more profound role for T cells in selecting for high-affinity antibody production than that of merely augmenting B-cell proliferation and thereby facilitating selection by antigen.

It has long been known that the magnitude of the immune response of old humans or experimental animals is reduced as compared with that of young animals. Several years ago, Goidl, Innes and Weksler[8] reported that old mice produce immune responses which are relatively lacking in high-affinity antibodies and IgG antibodies as compared to young mice. Shortly thereafter, in collaboration with Drs. Thorbecke, Goidl, Schrater and Kim, we began to examine anti-idiotype (id) antibody production. An idiotype is a set of antigenic determinants (idiotopes) located in or near the antigen-binding region of an antibody (AB_1 in this terminology) molecule. Anti-idiotype antibody (AB_2) is antibody specific for antigenic determinants located in or near the antigen-binding region of another antibody molecule. We used an enzyme-linked immunoadsorbent assay to detect serum anti-id[5] and hapten augmentation of plaque formation to detect auto-anti-id bound to B-cell surface id.[9,13,1] With the latter assay we obtained evidence that anti-id binding to B-cell surface Ig causes a reversible inhibition of antibody secretion. In the presence of a low concentration of hapten, the auto-anti-id is, in effect, displaced from the cell surface id; the cell begins to secrete antibody and can now be detected as a plaque-forming cell (PFC). Using both assays we obtained evidence that auto-anti-id is spontaneously produced following immunization of mice, rabbits and chickens with foreign antigens (TNP-BGG, TNP-Ficoll, etc.)[1,9,10,13] and that older animals produce increased amounts of auto-anti-id following primary immunization with these TNP-conjugates.[11] We suggested that this increased auto-anti-id production might be, at least in part, responsible for the decreased magnitude and affinity of the immune response of old mice.

To determine the cellular basis for these age-associated changes in the immune response, we used classical cell-transfer methods. Young irradiated mice reconstituted with spleen cells from naive old donors were shown to behave like old mice in that following immunization they produce a relatively low-affinity AB_1 response and a high auto-anti-id response.[12] In contrast, irradiated young mice reconstituted with bone marrow from old donors and thymus cells from young mice produce a young-like immune response characterized by high-affinity AB_1 and low auto-anti-id.[12] Thus, it appears that the bone marrow of old mice is capable of generating a young-like distribution of B-cell clones, and that it is the peripheral lymphoid population of the old mice which is responsible for their altered behavior. It was hypothesized that the long-lived peripheral T-cell population is responsible for these changes by shifting the distribution of peripheral B-cell clones through id–anti-id interactions, in the absence of antigen. To test this hypothesis, mice were reconstituted with anti-Thy-1-treated bone marrow from naive old or young mice together with purified splenic T cells from either young or old naive mice. When immunized three weeks after cell transfer, their immune response with respect to auto-anti-id production was characteristic of the age of the T-cell donor irrespective of the age of the B-cell donor.[16] This result could be due either to some T-cell helper effect occurring during the response to antigen or to a shift in B-cell id distribution due to id–anti-id interactions, between newly arising B cells and peripheral T cells, prior to antigen exposure. To distinguish these possibilities, serial cell transfers were conducted.[17] Irradiated mice were reconstituted with bone marrow from young naive donors together with purified splenic T cells from either young or old naive donors. The recipients were not immunized. Rather, after the cells had resided together for three weeks, the initial host was killed and its purified splenic B cells were transferred, together with T cells from young naive donors, into a second irradiated recipient which was immunized with TNP-BGG. The immune response of the second recipient, with respect to auto-anti-id production, was typical of the age of the T-cell donor to the first recipient. Since the first recipients were not exposed to antigen, the difference in their B-cell populations, as manifested in the second recipients, must reflect an antigen-independent shift in id distribution, presumably due to id–anti-id interactions between newly maturing B cells and peripheral T cells. Thus, the long-lived peripheral T-cell population appears capable of serving as a memory bank for shifts in clonal distribution occurring throughout life presumably due to interactions with external and self-antigens.

If the above reasoning were correct, then it would follow that, if old mice were treated so as to deplete their peripheral lymphoid system of lymphocytes and then were allowed to repopulate their peripheral lymphoid system with cells newly arising from their own bone marrow, then they would, after such auto-reconstitution, respond like young animals with respect to auto-anti-id production, id expression, and antibody affinity. This experiment can be accomplished by irradiating mice with a portion of their bone marrow shielded (so-called total nodal irradiation). Such irradiated animals repopulate their peripheral lymphoid organs and recover immune reactivity over a six- to seven-week period post-irradiation.[12] Old mice irradiated with partial bone-marrow shielding and immunized two months later produce a response more like that of a young animal, in that they have low auto-anti-id and relatively high Ab_1 affinity as well as an increased IgG to IgM ratio.[12,22] To further support the role of peripheral T cells in influencing B-cell clonal expression, young mice were irradiated with partial bone-marrow shielding and given purified splenic T cells from either young or old naive donors one day thereafter. Two months later they were immunized with TNP-BGG. Their immune response was typical of the age of the T-cell donor with respect to auto-anti-id and antibody affinity.[12] To determine whether these effects are due to an antigen-independent shift in B-cell clonal distribution or to helper T-cell activity operating during the response to antigen, cell transfers were carried out.[17] Young naive mice were irradiated with partial bone-marrow shielding and one day later received purified splenic T cells from either young or old naive donors. These primary recipients were not immunized. Two months after irradiation and T-cell transfer, they were killed and their purified splenic B cells were transferred into naive, young, irradiated recipients together with T cells from young donors. These second recipients were immunized one day after cell transfer and produced an auto-anti-id response typical of mice of the age of the T-cell donor to the primary recipient.[17] The results are thus consistent with the view that the T-cell influence on id expression is mediated via an antigen-independent shift in B-cell clonal distribution resulting from id–anti-id interactions between newly differentiating B cells and peripheral T cells.

Recent studies in our laboratory[18] and by Goidl[19] have indicated that, if auto-anti-id is eluted from the B cells of old immunized mice by brief incubation with hapten, one can then detect the secretion of high-affinity antibodies whose production cannot otherwise be detected. These results are consistent with the view that B cells potentially capable of producing high-affinity antibodies are present in old mice but are generally not detected because of their down-regulation by auto-anti-id. Studies by Zharhary and Klinman[25] have also suggested that the apparent decreased antibody diversity expressed in the immune response of old animals is due to regulatory effects operating on the B-cell population. Thus, the apparent reduced affinity of the immune response of old mice appears, at least in part, to be due to down-regulation by their marked auto-anti-id production. Why high-affinity antibodies are preferentially down-regulated by anti-id is not clear. However, it should be noted that antigen-antibody complexes appear to be more effective than antibody alone in stimulating anti-id production. High-affinity antibodies will form more stable complexes with antigen than do low-affinity antibodies. At low antigen concentration, antigen-antibody complexes would tend to involve mainly high-affinity antibodies which might lead to a preferential production of anti-id specific for ids on high-affinity antibody and account for the preferential depression of high-affinity antibody production in old mice.

In conclusion, there appears to be at least two distinct processes of selection operating on the B-cell population. One is antigen dependent and results in the increase in antibody affinity during the immune response and in the down-regulation of high-affinity antibody production during B-cell tolerance induction. The second is an antigen-independent shift in B-cell id distribution resulting from id–anti-id interactions between newly differentiating B cells and peripheral T cells. It is suggested that the long-lived peripheral T-cell population serves as a memory bank for life-long interactions with external and self-antigens whereby an optimally adaptive distribution of B-cell ids is maintained in the peripheral lymphoid system. Since anti-id production generally occurs during the immune response to external antigens, and since the auto-anti-id (and id specific T cells) so generated appears to be capable (depending upon conditions) of either stimulating or suppressing production of the corresponding id (and cross-reactive ids); it follows that during any immune response both antigen and anti-id serve as selective agents influencing the distribution of B cells. While previous modeling efforts have focused upon the role of antigen as a selective force leading to affinity maturation, it is likely that without considering the selective effects, both positive and negative, of anti-id on the evolving B-cell population, a fully satisfactory model of the antibody response cannot be developed.

REFERENCES

1. Bhogal, B. S., A. J. Edelman, J. J. Gibbons, E. B. Jacobson, G. W. Siskind, and G. J. Thorbecke. "Production of Auto-Anti-Idiotypic Antibody During the Normal Immune Response. X. Response to TNP-Ficoll in the Chicken." *Cell. Immunol.* **91** (1985):159–167.
2. DeKruyff, R., and G. W. Siskind. "Studies on the Control of Antibody Synthesis. XIV. Role of T Cells in Regulating Antibody Affinity." *Cell Immunol.* **47** (1979):134–142.
3. Eisen, H. N. "Equilibrium Dialysis for Measurement of Antibody-Hapten Affinities." *Methods Med. Res.* **10** (1964):106.
4. Eisen, H. N., and G. W. Siskind. "Variations in Affinities of Antibodies During the Immune Response." *Biochem.* **3** (1964):996–1008.
5. Gibbons, J. J., E. A. Goidl, G. M. Shepherd, G. J. Thorbecke, and G. W. Siskind. "Production of Auto-Anti-Idiotypic Antibody During the Normal Immune Response. XII. An Enzyme Linked Immunosorbent Assay for Auto-Anti-Idiotype Antibody." *J. Immunol. Methods* **79** (1985):231–237.
6. Goidl, E. A., J. J. Barondess, and G. W. Siskind. "Studies on the Control of Antibody Synthesis. VII. Changes in Affinity of Direct and Indirect Plaque Forming Cells with Time After Immunization in the Mouse: Loss of High Affinity Plaques Late After Immunization." *Immunology* **29** (1975):629–641.
7. Goidl, E. A., G. Birnbaum, and G. W. Siskind. "Determination of Antibody Avidity of the Cellular Level by the Plaque Inhibition Technique: Effect of Valence of Inhibitor." *J. Immunol. Methods* **8** (1975):47–51.
8. Goidl, E. A., J. B. Innes, and M. E. Weksler. "Immunological Studies of Aging. II. Loss of IgG and High Avidity Plaque-Forming Cells and Increased Suppressor Cell Activity in Aging Mice." *J. Exp. Med.* **144** (1976):1037–1048.
9. Goidl, E. A., A. F. Schrater, G. W. Siskind, and G. J. Thorbecke. "Production of Auto-Anti-Idiotypic Antibody During the Normal Immune Response to TNP-Ficoll. II. Hapten-Reversible Inhibition of Anti-TNP Plaque Forming Cells by Immune Serum as an Assay for Auto-Anti-Idiotypic Antibody." *J. Exp. Med.* **150** (1979):154–165.

10. Goidl, E. A., A. F. Schrater, G. J. Thorbecke, and G. W. Siskind. "Production of Auto-Anti-Idiotypic Antibody During the Normal Immune Response. IV. Studies of the Primary and Secondary Responses to Thymic Dependent and Independent Antigens." *Eur. J. Immunol.* **10** (1980):810–814.

11. Goidl, E. A., G. J. Thorbecke, M. E. Weksler, and G. W. Siskind. "Production of Auto-Anti-Idiotypic Antibody During the Normal Immune Response. V. Changes in the Auto-Anti-Idiotypic Antibody Response and the Idiotypic Repertoire Associated with Aging." *Proc. Natl. Acad. Sci. (USA)* **77** (1980):6788–6792.

12. Goidl, E. A., J. W. Choy, J. J. Gibbons, M. E. Weksler, G. J. Thorbecke, and G. W. Siskind. "Production of Auto-Anti-Idiotypic Antibody During the Normal Immune Response. VII. Analysis of the Cellular Basis for the Increased Auto-Anti-Idiotypic Production of Aged Mice." *J. Exp. Med.* **157** (1983):1635–1645.

13. Goidl, E. A., T. Hayama, G. M. Shepherd, G. W. Siskind, and G. J. Thorbecke. "Production of Auto-Anti-Idiotypic Antibody During the Normal Immune Response. VI. Hapten Augmentation of Plaque Formation and Hapten-Reversible Inhibition of Plaque Formation as Assays for Anti-Idiotype Antibody." *J. Immunol. Methods* **58** (1983):1–17.

14. Kim, Y. T., D. Greenbaum, P. Davis, S. A. Fink, T. P. Werblin, and G. W. Siskind. "Studies on the Control of Antibody Synthesis. IX. Effect of Boosting on Antibody Affinity." *J. Immunol.* **114** (1975):1302–1306.

15. Kim, Y. T., S. Kalvar, and G. W. Siskind. "A Comparison of the Farr Technique with Equilibrium Dialysis for Measurement of Antibody Concentration and Affinity." *J. Immunol. Methods* **6** (1975):347–354.

16. Kim, Y. T., E. A. Goidl, C. Samarut, M. E. Weksler, G. J. Thorbecke, and G. W. Siskind. "Bone Marrow Function. I. Peripheral T Cells are Responsible for the Increased Auto-Anti-Idiotype Response of Older Mice." *J. Exp. Med.* **161** (1985):1237–1242.

17. Kim, Y. T., T. DeBlasio, G. J. Thorbecke, M. E. Weksler, and G. W. Siskind. "Production of Auto-Anti-Idiotypic Antibody During the Normal Immune Response. XIV. Evidence for the Antigen Independent Operation of the Idiotype Network." *Immunol.* **67** (1989):191–196.

18. Marcenaro, L., C. Russo, Y. T. Kim, G. W. Siskind, and M. E. Weksler. "Immunologic Studies of Aging. Normal B-Cell Repertoire in Aged Mice: Studies at a Clonal Level." *Cell. Immunol.* **119** (1989):202–210.

19. Martin Mcvoy, S. J., and E. A. Goidl. "Studies on Immunological Maturation. II. The Absence of High-Affinity Producing Cells Early in the Immune Response is Only Apparent." *Aging: Immunol. Inf. Dis.* **1** (1988):47–54.

20. Siskind, G. w., and B. Benacerraf. "Cell Selection by Antigen in the Immune Response." *Advan. Immunol.* **10** (1969):1–50.

21. Theis, G. A., and G. W. Siskind. "Selection of Cell Population in Induction of Tolerance: Affinity of Antibody Formed in Partially Tolerant Rabbits." *J. Immunol.* **100** (1968):138–141.

22. Tsuda, T., Y. T. Kim, G. W. Siskind, and M. E. Weksler. "Old Mice Recover the Ability to Produce IgG and High Avidity Antibody Following Irradiation with Partial Bone Marrow Shielding." *Proc. Natl. Acad. Sci. (USA)* **85** (1988):1169–1173.
23. Werblin, T. P., and G. W. Siskind. "Distribution of Antibody Affinities: Technique of Measurement." *Immunochem.* **9** (1972):987–1011.
24. Werblin, T. P., Y. T. Kim, F. Quagliata, and G. W. Siskind. "Studies on the Control of Antibody Synthesis. III. Changes in Heterogeneity of Antibody Affinity During the Course of the Immune Response." *Immunology* **24** (1973):477–492.
25. Zharhary, D., and N. R. Klinman. "B Cell Repertoire Diversity to PR8 Virus Does Not Decrease with Age." *J. Immunol.* **133** (1984):2285–2287.

Richard G. Weinand
Department of Computer Science, Wayne State University, Detroit, Michigan 48202, U.S.A.

Somatic Mutation and the Antibody Repertoire: A Computational Model of Shape Space

INTRODUCTION

Somatic mutation has been implicated as a significant and possibly primary factor in affinity maturation and memory in the humoral immune response. When the immune system is challenged by antigen, stimulated B cells experience a hypermutation in the gene segments that code for the antigen-binding site of the antibody, creating new antibody specificities that were not present at the time of immunization. Although many of the mutations may be non-productive, new specificities with a higher affinity for the antigen are sometimes created. These higher-affinity cells are preferentially selected for proliferation and eventual antibody secretion, resulting in a progressively higher average affinity over time.

Informal models of the humoral immune response that incorporate somatic mutation have been presented by several authors, among them Berek et al.,[3] Manser et al.,[13] and Rajewsky et al.[18] Their models are descriptive in nature, and are derived from an analysis of nucleotide sequencing data obtained from experiments with immunized mice. These models are particularly important in that they clearly illustrate and emphasize the dynamic and stochastic nature of a continuously expanding and contracting repertoire of antibody specificities. In addition to providing some important new insights into the nature of the immune response, these models

Molecular Evolution on Rugged Landscapes, SFI Studies in the Sciences of
Complexity, vol. IX, Eds. A. Perelson and S. Kauffman, Addison-Wesley, 1991

and the studies on which they are based also raise some significant new questions that need to be addressed. The rate of somatic mutation has been estimated to be as high as one point mutation per cell division.[5,14,21] How are antibodies able to withstand such a barrage of random mutations and increase their affinity for the antigen? Are the resultant antibodies more specific or less specific for the antigen than their precursors? What impact might this have on the generation of memory cells? Random mutations also increase the possibility of generating antibodies that may be harmful to self. How is this risk controlled or minimized?

Our long-range objective is to develop a complete dynamical system model of the humoral immune response that can be used to investigate these kinds of questions. The first step in this process is to devise a representation for the antibody and antigen repertoires that is suitable for that purpose. Specifically, the representation must be at such a level that it permits the random mutation of an antibody and the subsequent calculation of its (perhaps) altered affinity for the antigen. This is a difficult modeling task because the representation must be sufficiently complex to capture the essential nature of antibody/antigen interactions, and at the same time enable the required calculations to be performed on a very large number of antibodies in a reasonable amount of time.

In this paper we present the design of such a representation, as well as an analysis of the characteristics of an antibody repertoire created by that design. We proceed by:

1. defining a set of behavioral characteristics that the repertoire is required to satisfy,
2. devising a three-dimensional representation for an antibody that permits the calculation of its affinity for an antigen as a function of its amino-acid sequence,
3. creating a number of artificial antibody/antigen repertoires based on this representation and testing their performance against the required behavioral characteristics, and
4. analysing some of the behavioral characteristics of the model repertoires when subjected to random somatic mutation.

REQUIRED CHARACTERISTICS OF THE ANTIBODY REPERTOIRE

The basic goal of our design is to create a model antibody repertoire that has, as a minimal requirement, properties and behavioral characteristics that are generally accepted as being representative of the real system. For the purpose of this model, then, the following characteristics are considered to be essential requirements of the model repertoires:

1. Most antigens react with B cells of more than one specificity, thereby initiating a response that is heterogeneous with regard to the affinity of antibodies produced.
2. There are more B cell specificities that have a low affinity for an antigen than those that have a high affinity for the antigen.
3. The substitution of a single base pair in the DNA sequence that codes for the antigen-binding site can have a wide range of effects on the binding properties of the antibody. The affinity can be either increased or decreased greatly, minimally, or not at al.[8,20]
4. The binding strength and specificity of an antibody generally exhibit an inverse relationship, i.e., antibodies with high average binding strengths are generally less specific than antibodies with low average binding strengths.[11,15]

DESIGN OF THE MODEL REPERTOIRE

The design of our model repertoire is derived from the concept of *shape space* introduced in Perelson and Oster,[16] and implemented in the form of a network model of the immune response in Perelson.[17] In this model the antibody/antigen, combining regions are represented as bit strings, and the strength of a bond is calculated as a function of the complementarity of the two strings. Our model repertoire is an extension of this basic idea.

The primary assumption behind our model repertoire is that the binding strength between an antibody and an antigen is a function of the complementarity of their molecular surfaces in terms of (a) size, (b) shape, and (c) functionality. Molecular size and shape determine the proximity of the functional sites, and their functional complementarity determines the magnitude of the bond.[2,4,7,19,22] This structural assumption is incorporated directly into our representation of the antibody and antigen repertoires.

PHYSICAL REPRESENTATION

The basic units of our model antibodies are amino acids, which are translated from nucleotide triplets (codons) according to a genetic code. We use a slightly modified version of the genetic code that maps the 64 possible codons onto 16 amino acids instead of 20 (Appendix I, Table 4). Each amino acid is composed of a number of 3-dimensional units arranged in a variety of shapes. Two examples of such amino acids are shown in Figure 1.

Each amino-acid shape has a common base four units in length. Attached to each of the base units are side extensions, each of which can vary in length from 0 to 3 units. This construction permits the formation of 256 (4^4) possible amino-acid configurations, any 16 of which define an antibody repertoire (Appendix II, Table 5).

The antigen-binding site of an antibody is represented as a 3-dimensional cavity one unit deep formed by the apposition of 4 amino acids (Figure 2).

The four amino acids are oriented on an 8 × 10 unit grid, with their bases aligned at the outer edges. This representation defines an antibody repertoire of size 65,536 (16^4). We will refer to a particular antibody in the repertoire by an ordered list of four letters, taken from the model genetic code, that identifies its amino-acid constituents. The antibody in Figure 2, for example, will be referred to as antibody [AHLR].

Antigenic epitopes are constructed of units similar to those of the antibodies, and are represented as projections from the surface of the antigen. Since there are many more possible antigens than antibodies (Inman[9] has estimated that there are at least 10^{16} distinguishable families of antigen sturctures), we allow an antigen to consist of any possible combination of units within an 8-unit-by-8-unit shape. The number of possible antigens that can be represented in this fashion is on the order of 10^{20}.

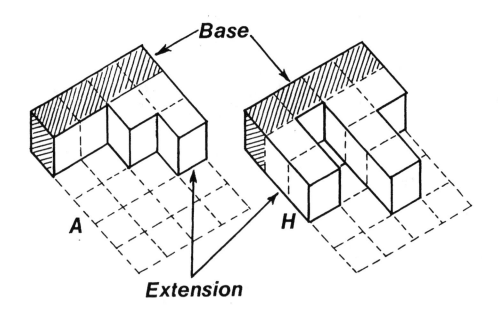

FIGURE 1   Configuration of amino acids A and H.

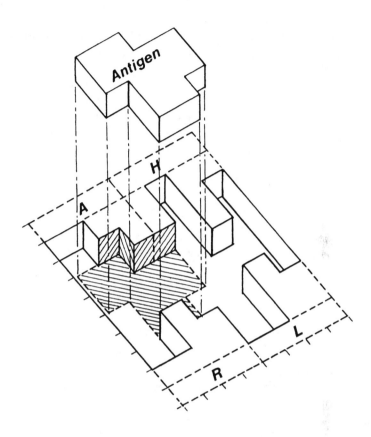

FIGURE 2   Antigen-binding site formed by amino acids A, H, L and R.

CALCULATION OF BINDING STRENGTH

The bond between an antigenic epitope and an antibody (expressed as an affinity constant) is the result of complex interactions between chemical functional groups on the surface of the antigen and amino-acid side groups in the binding site of the antibody. The types of forces involved include hydrogen bonds, electrostatic interactions, van der Waals interactions, and hydrophobic associations, where complementarity of size, shape, and functional groups is critical in forming and determining the strength of the bond. We make no attempt to mimic the real structures and forces involved in antigen-antibody binding, but utilize a much simplified representation that we think captures the essential features of the process. We also assume that for the purpose of calculating an affinity constant, antibody and antigen configurations can be treated as though they are rigid and do not undergo conformational change upon binding.[1,19]

Each unit surface that forms a portion of the floor or walls of the cavity is an accessible contact surface for an antigen, and is assigned one of five functional types. These functional types are somewhat arbitrary in the sense that they do not correspond directly with real functional types, but provide a means for representing molecular functionality as an important parameter in antibody-antigen interactions. The binding strength between an antibody and an antigen is calculated as a function of the number and type of functional surfaces that make contact with each other, according to the following rules:

1. An epitope must "fit" completely within the cavity to form a bond.
2. A force is exerted only where two unit surfaces come into full contact.
3. The force exerted at each contact location is determined by the functional types of the two contacting surfaces.
4. The total bond between an antigen and an antibody is defined by some function of the individual forces.
5. The binding force is considered to the "best fit" force, considering all possible unit translations and 90° rotations of the two shapes.

The binding force between any two types of functional units that make contact is defined by a matrix, an example of which is shown in Table 1.

Each value in the body of Table 1 represents the units of binding force exerted between a pair of functional types on the antibody and antigen surfaces that make contact. For example, if a unit surface of functional type c on an antibody makes contact with a unit surface of functional type b on an antigen, the strength of that particular bond would be 6. A major feature of this representation is that it permits us to easily vary the relative importance of the three major binding strength parameters: size, shape, and functionality. The binding characteristics of

TABLE 1 Binding Force Matrix: Functional Type of Contacting Surface

			antigen		
antibody	a	b	c	d	e
a	1.0	1.0	1.0	1.0	1.0
b	1.0	-4.0	6.0	1.0	1.0
c	1.0	6.0	-4.0	1.0	1.0
d	1.0	1.0	1.0	4.0	1.0
e	1.0	1.0	1.0	1.0	11.0

the repertoire can be varied by assigning different functional types to each of the units that make up the amino-acid and antigenic surfaces, and by assigning various values to the weights in the function matrix.

The affinity constant for an antibody-antigen combination is calculated by first summing the individual bonds for each of the unit surfaces making contact. This produces a total bond expressed as a number of arbitrary bond units, generally falling in the range from 0 to 100. The total bond units are then mapped onto a physiologically relevant range of affinity constants by the function given in Eq. (1).

$$K_{ij} = c_1 c_2^{B_{ij}}, \qquad (1)$$

where K_{ij} = affinity constant for antibody i, antigen j,
$\quad B_{ij}$ = total bond units for antibody i, antigen j,
$\quad c_1$ = scaling factor, and
$\quad c_2$ = slope factor.

This function permits us to control the number and distribution of antibody affinities that will be triggered by an antigen by varying the threshold affinity and the values of c_1 and c_2. Throughout the remainder of this paper we will use the terms "affinity" and "bond strength" interchangeably.

ANALYSIS OF THE MODEL REPERTOIRE

Four complete antibody repertoires of 65,536 antibodies each (Repertoires A, B, C, and D), were generated along with a random sample of 100 antigen configurations. In this section we analyze the binding characteristics of these repertoires and compare them with the requirements formulated previously.

REPERTOIRE A

The characteristics of the generated antibody/antigen repertoires with regard to the requirements for (1) heterogeneity of affinity, and (2) distribution of affinities can be seen in Figure 3.

Figure 3 is a binding distribution for antigen #84, an "average" antigen from the sample of 100. Antigen #84 has a bond strength greater than zero for 4,098 antibody configurations in the potential repertoire of 65,536 (6.25%). The distribution of bond strengths is roughly symmetrical, ranging from a low of 2 to a high of 82. If we consider a bond strength of 65 ($K_a \approx 1.0 \times 10^5 M^{-1}$) to be the minimum required for triggering an immunogenic response, then 90 of the antibodies (0.137% of the complete repertoire) would be *triggerable*. The proportion of triggerable antibodies is also seen to be a generally decreasing function of affinity, i.e., there are many more antibodies of low affinity than high affinity. These characteristics satisfy the first two requirements.

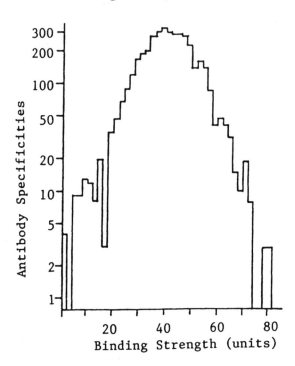

Binding Strength Distribution

FIGURE 3   Histogram showing the distribution of binding strengths of antibody repertoire A vs. antigen #84. The number of antibody specificities (ordinate) is plotted against binding strength units (abcissa).

The third requirement of the model repertoire is that the substitution of a single amino acid should be capable of producing a wide range of effects on the affinity of the antibody. We evaluated this property of the model repertoire by deriving the set of all possible antibodies that could be generated by the substitution of a single amino acid for each of the 90 triggerable antibodies. As an example, in Table 2 we see that antibody [EPIH] can generate 27 other antibodies.

Based on the model genetic code (Appendix I, Table 4), there are six possible substitutions for amino acid E in position 1, seven for amino acid P in position 2, eight for amino acid I in position 3, and six for amino acid H in position 4. Four of these mutant antibodies (indicated by "*") are *productive*, i.e., they have a bond strength at least as great as the triggering threshold of 65 bond units. Antibody [KPIH] has the same bond strength of 70 as antibody [EPIH], antibodies [EPIP] and [DPIH] have higher bond strengths of 71 and 79 respectively, and antibody [VPIH] has a lower bond strength of 65. We also note that amino acids P and I in positions

2 and 3 respectively are absolutely essential to the formation of a productive bond. All substitutions of either of these amino acids are non-productive. At the other extreme, amino acid E in position 1 is productively replaceable 50% of the time.

Many of the other triggerable antibodies are similar to antibody [EPIH] in that they can generate a number of productive mutations, some of which have a higher affinity and some of which have the same or lower affinity. There are also several other antibodies, such as [IIPD], that cannot generate a single productive antibody. The substitution of any one of their amino acids results in the total loss of their ability to be triggered by antigen #84.

The mutability properties of the antibody repertoire are considered from a slightly different perspective in Kauffman et al.,[12] where the authors discuss somatic mutation in the context of an *affinity landscape*. One of the important properties of the landscape is the degree of correlation between the affinity of each antibody and the affinities of its one-mutant neighbors (those antibodies differing by a single amino acid). A completely uncorrelated landscape would be one where the affinity of one antibody carries no information about the affinities of its one-mutant neighbors. Although the actual degree of correlation of the real antibody landscape is unknown, we would expect that it is correlated to some degree. In other words, if a given antibody has high affinity for an epitope, we might expect a large fraction of its one-mutant neighbors to also have high affinity for that epitope. We investigated this property for the model repertoire by deriving a number of statistics for the set of triggerable antibodies as listed in Table 3.

TABLE 2 Possible mutations of antibody [EPIH][1]

Position 1		Position 2		Position 3		Position 4	
Antibody	Bond	Antibody	Bond	Antibody	Bond	Antibody	Bond
[A---]	54	[-A--]	0	[--V-]	0	[---L]	57
[V---]*	65	[-L--]	51	[--L-]	0	[---P]*	71
[G---]	0	[-S--]	0	[--M-]	0	[---N]	0
[N---]	0	[-T--]	0	[--S-]	0	[---C]	0
[D---]*	79	[-N--]	0	[--T-]	0	[---D]	0
[K---]*	70	[-R--]	0	[--C-]	33	[---R]	0
		[-H--]	41	[--K-]	0		
				[--R-]	0		

[1] Unchanged amino acids are indicated by a "-"; productive mutations are indicated by an "*".

TABLE 3 Characteristics of Triggerable Antibodies

bond group (units)	no. of antibody specificities	no. of productive mutants	avg. bond strength of mutants	avg. no. of productive mutants
65–67	46	109	67.41	2.37
68–70	11	30	72.47	2.73
71–73	22	77	72.36	3.50
74–76	5	15	71.67	3.00
77–79	1	5	69.80	5.00
80–82	5	26	73.52	5.20
All	90	262	70.35	2.91

The data in Table 3 reveal two important characteristics of the model repertoire. The first is that there is little correlation between the bond strength of an antibody and the average bond strength of its productive mutants, with the exception perhaps of the two extremes. The average bond strength of the mutants arising from the lowest-affinity group is significantly lower than the overall average, and the average bond strength of the mutants arising from the highest-affinity group is significantly higher than the overall average. It must be kept in mind, however, that these data are based on productive mutants only, i.e., those with an affinity greater than or equal to the triggering threshold. Although the inclusion of the below-threshold affinity mutants might alter this result, we chose not to include them here because we are interested only in the characteristics of the productive antibodies at this time.

The second point is that there is a definite and strong correlation between the affinity of an antibody and the average number of productive mutants that it can generate. Although the average affinity of the productive mutants generated by high-affinity antibodies is not much different than that for low-affinity antibodies, high-affinity antibodies generate more mutants which survive and continue to proliferate and mutate. This is an important property of the repertoire, which we will discuss in more detail later on when we analyze the effects of somatic mutation on clonal development.

The fourth requirement of the model repertoire concerns the relationship between specificity and affinity. Antibodies vary greatly in the number of antigens that they can bind as well as in the strengths of their bonds. Using a random sample of 61 antigens, we calculated the average binding strength of each antibody in the repertoire for each of the antigens. These average binding strengths were grouped into categories based on the number of antigens bound, and the results are displayed in Figure 4.

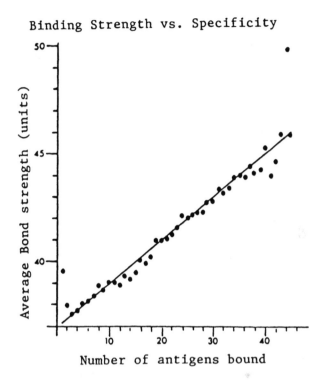

FIGURE 4 Plot of the relation between affinity and specificity. Each data point represents the average bond strength (ordinate) of all antibody specificities that bind a specific number (abcissa) of different antigens.

These data clearly exhibit the inverse relationship between specificity and affinity. Antibodies with a high average bond strength (affinity) are seen to be generally less specific (bind more different antigens) than antibodies with a low average affinity. It must be pointed out, however, that this is a general relationship. Exceptions where an antibody has both high affinity and high specificity do occasionally occur in the model repertoire.

REPERTOIRES B, C, AND D

Three additional repertoires (B, C, and D) were generated and evaluated to test the generality of the results presented above. Repertoire B was created by randomly selecting a different set of amino acids as its base, leaving the values in the function matrix unchanged. Repertoires C and D were then created by using the amino acids of Repertoires A and B, and modifying the values in the function matrix to give less weight to the functionality parameter. The behavioral characteristics of

Repertoires B, C, and D are found to be essentially the same as those of Repertoire A. The major differences in the four repertoires are in the range and variability of their binding strength distributions. Repertoires A and B, which have high values assigned to functionality, generate distributions that cover a wide range of affinities and are somewhat discontinuous, containing a number of significant peaks and valleys. Repertoires C and D, on the other hand, produce binding strength distributions that span a much narrower range of binding strengths, and are also more continuous than those of Repertoires A and B.

EFFECT OF SOMATIC MUTATION ON REPERTOIRE DIVERSITY

It is beyond the scope of this initial investigation to test the antibody repertoire model in a dynamical system model of the immune response. We can, however, extend our present analysis of the model repertoire by considering at least some of the major effects of somatic mutation on generation of antibody diversity.

The process by which somatic mutation generates antibodies with increased affinity for an antigen has been demonstrated to occur in a step-wise fashion.[5,21] New antibody specificities are generated primarily by single point mutations, and those mutants with a greater-than-threshold affinity for the antigen (*productive* antibodies) are selected for continued proliferation and mutation. Mutant antibodies with lower than threshold affinity for the antigen (*non-productive* antibodies), on the other hand, are excluded from further participation. An important consequence of this process is that new specificities of productive antibodies can only be generated from mutations of other productive antibodies that are *already present in the system*. If the sequential mutation path from one productive antibody to another includes one or more non-productive antibodies, that path is effectively blocked. B cells that express non-productive antibodies are excluded from the proliferating pool and therefore do not generate additional mutations.

We explored this characteristic of the antibody repertoire by deriving and analyzing the set of all possible productive antibodies that could arise from antibody [VIIE] as the result of a sequence of one or more point mutations (Figure 5).

Antibody [VIIE] is seen to be a member of a connected set of eight antibodies, which we term a *mutation set*. Any antibody in the set can be reached from any other antibody in the set by either a single point mutation or a sequence of point mutations because each of the antibodies in the set is productive. Mutations to antibodies not contained in this set do occur, but lead to an immediate dead end because they are not of sufficiently high affinity to remain in the proliferating pool. By direct inference, then, we can also see that (a) none of the other 82 productive antibodies in the repertoire can be reached from any of the antibodies in this set, and (b) mutations *into* this set from antibodies outside the set are not possible.

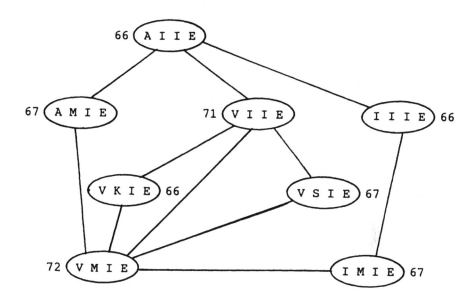

FIGURE 5   Diagram of the set of productive antibody specificities derivable from antibody [VIIE] by a step-wise sequence of single amino-acid substitutions permitted by the genetic code.

By definition, antibodies that are directly mutable into each other differ by one amino acid. We might expect, then, to see a great deal of structural similarity between the antibodies in a given mutation set. In the mutation set of Figure 5, the structural similarity is evidenced by the fact that each antibody in the set has the same amino acids I and E in positions 2 and 3. We also note that the amino acids in these two positions are essential to the formation of a productive bond since all mutations in positions 2 and 3 generate non-productive antibodies. Any change in either of them reduces the affinity of the antibody below the triggering threshold. Structural similarity between antibodies in a mutation set is not a necessary requirement in general, however. In principle, two antibodies that are separated by four mutations could each be composed of four completely different amino acids and be quite dissimilar structurally. The likelihood of occurrence, however, would be quite low.

The mutation set of another productive antibody [VPIH] is depicted in Figure 6.

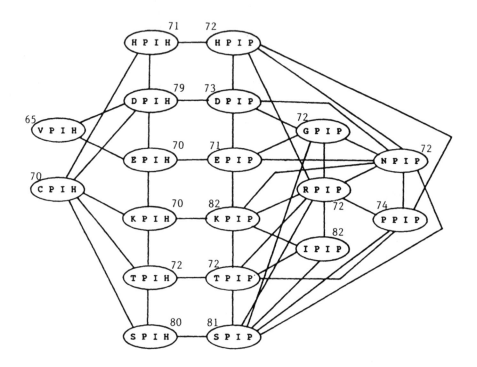

FIGURE 6 Diagram of the set of productive antibody specificities derivable from antibody [VPIH] by a step-wise sequence of single amino acid substitutions permitted by the genetic code.

This mutation set has 19 antibodies, and has characteristics similar to those of the set previously described. The antibodies in this set also have a great deal of structural similarity. Each antibody has the same amino acids (P and I) in positions 2 and 3, 11 of the antibodies have amino acid P in position 4, and the other 8 have amino acid H in that position. The complete set of 90 productive antibodies, grouped by mutation set, are summarized in Appendix III (Table 6).

We return, here, to the idea of an affinity landscape.[12] The landscape contains a number of local optima which are antibodies that have a higher affinity than all of their one-mutant neighbors. In this view, maturation of the immune response via somatic mutation is seen as a set of adaptive walks on the landscape from the initial germline specificities through a succession of one-mutant neighbors of successively higher affinity, toward one or more of the local optima. When steps along the walk are allowed only in the direction of higher affinity, the reaching of a local optima prevents further exploration of the potential repertoire from that point on the landscape. An example of this is depicted in Figure 6, which contains three local optima: antibody [DPIH] with a bond strength of 79, and antibodies [KPIP] and [IPIP], both with a bond strength of 82. Antibody [DPIH] is prevented from mutating to either of the two higher-affinity local optima because it would have to move through mutations of lower affinity.

In the real system, antibody mutants of lower affinity remain in the proliferating/mutating pool *as long as they are of sufficiently high affinity to be triggered by the antigen.* Since all antibodies in a mutation set are productive (by definition), the highest-affinity antibody in a mutation set can always be reached from any other antibody in the set by a sequence of one or more point mutations. In this revised sense, then, each mutation set can be considered as having one and only one local optimum (not counting ties).

The partitioning of all productive antibodies of the potential repertoire into a number of disjoint mutation sets can have important consequences with regard to the total diversity that can be generated by somatic mutation. On the one hand, mutation sets ensure that the highest-affinity antibody in a mutation set can be reached by somatic mutation, *once the set has been activated.* By activation we mean that at least one antibody in the set has been triggered by the antigen. On the other hand, the initial activation of a particular mutation set is not generally assured, particularly if the mutation set contains few antibodies. The only mutation sets that will be activated are those that contain one or more antibodies that happen to be present at the time of immunization. Since the number of antibody specificities that are initially present is probably at least several orders of magnitude smaller than the size of the potential repertoire, it is quite likely that many if not most of the mutation sets will not be activated. Additional mutational sets may also be activated by the random synthesis of new B cells as the response progresses, but this is not likely to be a significant factor. The major point to be made here is that somatic mutation would appear to provide an effective means for the rapid exploration of a *limited* segment of the potential repertoire.

SUMMARY
SUITABILITY OF THE MODEL REPERTOIRE

The major objective of our initial investigation has been to devise a model repertoire that is suitable for investigating the process of somatic mutation in a full system model of the humoral immune response. We have presented the design of such a repertoire, analyzed its characteristics in relation to a set of pre-defined requirements, and also explored some of its characteristics under the influence of somatic mutation. Although our analysis does not represent a complete and systematic exploration of all the properties and characteristics of the model repertoire, we conclude that is adequate for the intended purpose, and that our major objective has been met.

LIMITATION OF THE MODEL REPERTOIRE

Our analysis of the properties and characteristics of the model repertoire has also helped to identify some of the limitations inherent in the representation we have chosen. A major limitation involves the important property of *gradualism*, which we discuss briefly below.

X-ray crystallographic studies have shown that the binding site of an antibody appears as an irregularly shaped surface formed primarily by the amino acids that constitute the complementarity-determining regions (CDRs) of the variable heavy (V_H) and light (V_L) chains.[1,2] Although the CDRs consist of about 60 amino acids, only a portion of them are able to make direct contact with the antigen due to the limited size of the binding site.[10] The study by Alzari et al,[1] for example, shows that seventeen amino acids from the antibody make contact with the antigenic epitope. Fifteen of the amino acids are in the CDRs and two are in the framework regions. An important property of this type of structure is that it permits gradual (as well as drastic) changes in shape and function to result from the substitution of a single amino acid. Substitutions of amino acids that form the surface of the binding site are most likely to produce large increases or decreases in affinity, whereas substitutions of other amino acids are more likely to produce more subtle effects, perhaps by inducing slight conformational changes in the binding sites. Our model antibodies do not have this important property of gradualism that is essential in evolutionary systems.[6] They consist of V regions that have only four amino acids (all of which directly form the antigen binding site), and have no framework or constant regions. All amino-acid substitutions, therefore, are likely to produce significant changes in the binding strength because one-fourth of the binding surface is changed.

If the gradualism feature were to be included in the current model, the computational requirements would be vastly increased since it requires the inclusion of many additional amino acids in the representation of an antibody. On the positive side, we have also gained some important insights regarding alternative methods for modeling the process of somatic mutation that should permit us to overcome this limitation. Our future efforts will take advantage of those insights.

THE CONCEPT OF MUTATION SETS

Although the dynamics of somatic mutation can only be fully explored in a complete system model, our analysis allows us to make the following prediction. Somatic mutation, although random, provides a controlled mechanism for expanding the pre-immune repertoire into a *limited* segment of the potential repertoire as follows:

a. The genetic code and the structure of the antibody repertoire, in combination with threshold affinity-based selection by antigen, dynamically partitions all triggerable clones in the potential repertoire into a number of mutation sets.

b. A limited number of mutation sets are *activated* when the system is challenged by an antigen, i.e., those mutation sets containing one or more antibodies that are *present* and *triggerable* by the antigen.

c. The complete set of triggerable antibody specificities in the potential repertoire that can be reached by somatic mutation is defined by and limited to those antibodies belonging to the activated mutation sets.

Although this conclusion was derived from the analysis of data generated by an artificial antibody repertoire, we believe it can be generalized to the real system as well. The concept of mutation sets can be logically derived for the real system from a few basic assumptions, independent of the representation chosen for our model repertoire. The size of mutation sets in the real system may very well be much larger (or even smaller) than in our model system, but we expect that even very large size differences would not change the nature of the results obtained. Their significance, however, may be altered.

APPENDIX

TABLE 4 Model Genetic Code (16 amino acids)[1]

a.a. no.	standard codes	model code	nucleotide triplets	no. of triplets
	(hydrophobic)			
1	A ala	A	GC*	4
2	V val	V	GU*	4
3	L leu	L	CU*, UU{A,G}	6
4	I ile	I	AU{U,C,A}	3
5	P pro	P	CC*	4
6	M met	M	AUG	1
7	F phe	M (#)	UU{U,C}	2
8	W trp	M (#)	UGG	1
	(hydrophilic)			
9	G gly	G	GG*	4
10	S ser	S	UC*, AG{U,C}	6
11	T thr	T	AC*	4
12	N asn	N	AA{U,C}	2
13	C cys	C	UG{U,C}	2
14	Y try	C (#)	UA{U,C}	2
15	Q gln	C (#)	CA{A,G}	2
	(negatively charged)			
16	D asp	D	GA{U,C}	2
17	E glu	E	GA{A,G}	2
	(positively charged)			
18	K lys	K	AA{A,G}	2
19	R arg	R	CG*, AG{A,G}	6
20	H his	H	CA{U,C}	2
	(terminal)		UA{A,G}, UGA	3
			total	64

[1] * = {A,C,G,U}; # = deviations of the model code from the standard code.

TABLE 5 II. Amino Acid Shapes
(Repertoire A)

TABLE 6 III. List of All Triggerable Clones in Repertoire A (by Mutation Set)

Mutation Set No.	No. of Clones in Set	List of Clones
1	4	[RLPL]-66, [RIPL]-66, [RDPL]-69, [RHPL]-66
2	1	[ADPL]-65
3	3	[IPPI]-71, [SPII]-70, [KPII]-71
4	4	[IIIP]-67, [SIIP]-68, [KIIP]-67, [SIIH]-65
5	19	[IPIP]-82, [PPIP]-74, [GPIP]-72, [SPIP]-81, [TPIP]-72, [NPIP]-72, [DPIP]-73, [EPIP]-71, [KPIP]-82, [RPIP]-72, [HPIP]-72, [VPIH]-65, [SPIH]-80, [TPIH]-72, [CPIH]-70, [DPIH]-79, [EPIH]-70, [KPIH]-70, [HPIH]-71,
6	7	[VMIP]-72, [AMIH]-69, [VMIH]-80, [LMIH]-70, [IMIH-70], [MMIH]-70, [VKIH]-72
7	7	[IGIP]-65, [KGIP]-65, [IEIP]-74, [PEIP]-66, [SEIP]-73, [KEIP]-74, [SEIH]-72
8	2	[DEIP]-65, [DEIM]-71
9	9	[IPDP]-67, [SPDP]-66, [KPDP]-67, [IPEP]-74, [PPEP]-73, [SPEP]-73, [KPEP]-74, [SPDH]-65, [SPEH]-72
10	2	[DPEP]-62, [DPEH]-71
11	3	[IEEP]-66, [SEEP]-65, [KEEP]-66
12	1	[IIPD]-67
13	1	[IPPD]-67
14	1	[MDPD]-66
15	1	[IEPD]-70
16	1	[IAMD]-67
17	14	[MPIE]-67, [GPIE]-65, [TPIE]-65, [NPIE]-66, [DPIE]-66, [KPIE]-68, [RPIE]-66, [HPIE]-67, [MLIE]-66, [NLIE]-65, [DLIE]-66, [KLIE]-67, [RLIE]-65, [HLIE]-66
18	8	[AIIE]-66, [VIIE]-71, [IIIE]-66, [AMIE]-67, [VMIE]-72, [IMIE]-67, [VSIE]-67, [VKIE]-66
19	1	[VMDH]-65
20	1	[VMLH]-72
	90	

REFERENCES

1. Alzari, P. M., M. B. Lascombe and R. J. Poljak. "Three-Dimensional Structure of Antibodies." *Ann. Rev. Immunol.* **6** (1988):555–80.
2. Amit, A. G., R. A., Mariuzza, S. E. V. Phillips, and R. J. Poljak. "Three-Dimensional Structure of an Antigen-Antibody Complex at 2.8 A Resolution." *Science* **233** (1986):747–53.
3. Berek, C., and C. Milstein. "Mutation Drift and Repertoire Shift in the Maturation of the Immune Response." *Immunol. Rev.* **96** (1987):23–41.
4. Chothia, C., and J. Janin. "Principles of Protein-Protein Recognition." *Nature* **256** (1975):705–708.
5. Clarke, S. H., K. Huppi, D. Ruezinsky, L. Staudt, W. Gerhard, and M. Weigert. "Inter- and Intraclonal Diversity in the Antibody Response to Influenza Hemagglutinin." *J. Exp. Med.* **161** (1985):687–704.
6. Conrad, M. *Adaptability: The Significance of Variability from Molecule to Ecosystem.* New York: Plenum Press, 1983, 191–197.
7. Geysen, H. M., J. A. Tainer, S. J. Rodda, T. J. Mason, H. Alexander, E. D. Getzoff, and R. A. Lerner. "Chemistry of Antibody Binding to a Protein." *Science* **235** (1987):1184–1190.
8. Griffiths, G. M., C. Berek M. Kaartinen, and C. Milstein. "Somatic Mutation and the Maturation of Immune Response to 2-Phenyl Oxazolone." *Nature* **312** (1984):271–275.
9. Inman, J. K. "The Antibody Combining Region: Speculations on the Hypothesis of General Multispecificity." *Theoretical Immunology*, edited by G. I. Bell and A. S. Perelson. New York: Dekker, 1978, 243–278.
10. Kabat, E. A. *Structural Concepts in Immunology and Immunnochemistry,* second edition. New York: Holt, Rinehart and Winston, 1976.
11. Karush, F. "The Affinity of Antibody: Range, Variability and the Role of Multivalence, in Comprehensive Immunology." *Immunoglobulins*, Vol. 5, edited by G. W. Litman and R. A. Good. New York: Plenum, 1978, 85–116.
12. Kauffman, S. A., E. D. Weinberger, and A. S. Perelson. "Maturation of the Immune Response via Adaptive Walks On Affinity Landscapes." In *Theoretical Immunology, Part One*, edited by A. S. Perelson. Santa Fe Institute Studies in the Sciences of Complexity, Proc. Vol. II. Redwood City: CA: Addison-Wesley, 1988, 349–382.
13. Manser, T., L. J. Wysocki,M. N. Nargolies, and M. L. Gefter. "Evolution of Antibody Variable Region Structure During the Immune Response." *Immunol. Rev.* **96** (1987):142–162.
14. McKean, D., K. Huppi, M. Bell, L. Staudt, W. Gerhard, and M. Weigert. "Generation of Antibody Diversity in the Immune Response of BALB/c Mice to Influenze Virus Hemagglutinin." *Proc. Natl. Acad. Sci. USA* **81** (1984):3180–3184.

15. Ninio, J. "Adaptation and Evolution in the Immune System." In *Evolutionary Processes and Theory*, edited by S. Karlin and E. Nevo. New York: Academic Press, 1986, 143–165.

16. Perelson, A. S., and G. F. Oster. "Theoretical Studies of Clonal Selection: Minimal Antibody Repertoire Size and Reliability of Self- Non-Self Discrimination." *J. Theoret. Biol. 81* (1979):645–670.

17. Perelson, A. S. "Toward a Realistic Model of the Immune System." In *Theoretical Immunology*, Part Two, edited by A. S. Perelson. Santa Fe Institute Studies in the Sciences of Complexity, Proc. Vol. II. Redwood City, CA: Addison-Wesley, 1988, 377–401.

18. Rajewsky, K., I. Forster, and A. Cumano. "Evolutionary and Somatic Selection of the Antibody Repertoire in the Mouse." *Science* **238** (1987):1088–1094.

19. Rebek Jr., J. "Model Studies in Molecular Recognition." *Science* **235** (1987):1478–1484.

20. Rudikoff, S., A. M. Giusti,W. D. Cook, and M. D. Scharff. "Single Amino Acid Substitution Altering Antigen-Binding Specificity." *Proc. Natl. Acad. Sci. USA* (1982):1979–83.

21. Sablitzky, F., G. Wildner, and K. Rajewsky. "Somatic Mutation and Clonal Expansion of B Cells in an Antigen-Driven Immune Response." *EMBO J.* **4** (1985):345–350.

22. Sela, M. "Antigenicity: Some Molecular Aspects." *Science* **166** (1969):1365–1374.

Applied Molecular Evolution

Wlodek Mandecki
Corporate Molecular Biology D93D, Abbott Laboratories, Abbott Park, IL 60048, ph. (312) 937-2236

Evolution of Proteins from Random Sequences: A Model for the Protein Sequence Space and an Experimental Approach

Recent experimental data on the activities of a large number of single and multiple mutants of several proteins are interpreted from the perspective of general properties of the protein sequence space. A simple model is proposed for the sequence space. The observed frequencies of overlapping genes are then used to derive an estimate for the probability of obtaining a functional protein from a large pool of random sequences. In the second part of the paper a method is described for construction of long, randomized, open reading frames of nucleic acid sequences coding for proteins to provide an experimental approach for verification of the model.

INTRODUCTION

A critical question in the molecular evolution of proteins is how the ensemble of proteins currently existing in nature was selected from the potential pool of protein sequences (10^{400} sequences 300 a.a. in length). It can be estimated that the number of protein molecules which have existed in nature throughout the history of life on Earth is on the order of 10^{40}–10^{50} sequences. Therefore, only a minute fraction of the potentially available sequences could have been explored for function in biological systems.

Molecular Evolution on Rugged Landscapes, SFI Studies in the Sciences of Complexity, vol. IX, Eds. A. Perelson and S. Kauffman, Addison-Wesley, 1991

239

A hypothesis has been proposed that functional proteins can be selected from a pool of random sequences.[37] The selected proteins are then memorized and edited to provide evolved proteins and enzymes. This hypothesis brings up the question of what is the minimal number of random sequences in a pool sufficient for obtaining a functional protein. Below, this issue is addressed by both theoretical considerations and modeling, as well as an experimental approach.

There is experimental evidence that short random DNA or protein sequences can have a biological function. Prokaryotic promoters[14,40] or ribosome binding sites[34] have been shown to be selectable from random sequences. A large fraction of random signal sequences promote protein secretion.[1,17] Apparently, the only constraint on the sequences of activator domains of some regulatory proteins is a high content of acidic residues.[28] Thermal polymers of amino acids, protenoids, were shown to have a variety of enzymatic activities.[6] In addition, it becomes clear that proteins can tolerate a significant degree of sequence changes. It was found for a few protein systems on which extensive mutagenesis was performed that from 15% to 70% of point mutations do not recognizably affect the biological activity[16,23,33,38] of the protein. Moreover, replacements of contiguous[4,15] or non-contiguous[27] 3–5 amino-acid residues in several proteins resulted in a creation of active enzymes at a rate of 0.01% to 5% of sequences tested, even if the mutated residues are in the immediate vicinity of the catalytic residue.[3] In some cases, the replacement resulted in an increase of the specific activity of the enzyme.[4] The above suggests that the number of random polypeptides necessary for selection of a biologically active protein may be within the reach of modern molecular biology techniques, at least for the simplest protein functions.

A MODEL FOR THE PROTEIN SEQUENCE SPACE
OUTLINE

Mutagenesis experiments rely on the construction of point mutations, clusters of point mutations, deletions and insertions, or combinations of some or all of these to test properties of genes or proteins. The results of these experiments are highly variable, and generally the properties and activity of the mutant proteins generated depend on such factors as the protein itself, position of the mutation, the amino acid substituted, assay conditions, etc. Such results clearly reflect the complexity of protein folding and the active site. However, a question can be raised as to whether, despite such complexity, some general patterns of mutagenesis can be discerned.

An attempt is made below to derive the likelihood of reducing, maintaining, or increasing the fitness of a protein as a result of point mutation(s). Examples of fitness functions (f-f) are: the turnover number of an enzyme (k_{cat}), the association/dissociation constants, temperature stability, and other physico-chemical properties. It will be shown how such an approach can be used to rationalize the results of several mutagenesis experiments, as well as to aid in the design of mutagenesis strategies to improve properties of proteins. A simple model for the protein fitness space will be formulated.

The observed frequencies of overlapping genes[7,36] (encoded by the same DNA sequence, but translated in two different reading frames) are used to derive from the model an estimate for the probability of obtaining a functional protein from a large pool of random sequences. One can reason that if the evolution of proteins from non-optimized or random sequences was an insurmountable problem, there would be no (or very few) overlapping genes. Alternatively, if the evolution of proteins from nonoptimized sequences was easily achievable, overlapping genes would be abundant. Since overlapping genes occur at a measurable frequency, this frequency can in principle be the basis for calculating the probability of the occurrence of a functional protein among nonoptimized (random) sequences.

The concept of the sequence space was introduced by Hamming,[11] and its application to the protein sequence space was proposed by Maynard-Smith.[32] The distance function in the sequence space is defined as the number of amino acid (a.a.) differences between two protein sequences (Hamming distance). The number of the possible mutants that scan the closest neighborhood of a given protein sequence (i.e., within the distance of unity) is therefore $19 \times N$, where N is the length of a protein (in a.a.). A scalar function over the sequence space defines a fitness space. Such function typically provides a quantitative characterization of suitability of a protein sequence (here: point in the sequence space) to perform a specific function. Successive introduction of several point mutations into a given protein sequence defines a walk in the fitness space. A walk in the fitness space can represent, for instance, the evolution of a protein. A model for adaptive walks in the fitness space was developed by Kauffman et al.[19,20] The purpose of the analysis in this paper is to provide a simple model for the protein fitness space, which would be consistent with the results of mutagenesis experiments aimed at the construction of a large set of mutants.

POINT MUTATIONS

Below, data will be presented to provide the empirical basis for deriving probabilities of a silent mutation, and mutations that decrease, or increase, the fitness function value (down-mutation, or up-mutation, respectively).

Several studies have determined the frequency of point mutations, which would not significantly change the catalytic or binding properties of the given protein. Miller et al. used a collection of *E. coli* suppressor strains to introduce 1,500 missense mutations at sites corresponding to a stop codon in the *lac* repressor,[22] as well as 300 mutations in thymidilate synthetase.[21] Thirteen substitutions per site were made. The study showed that about 50% of the mutants had levels of activity indistinguishable from that of the wild type. Forty-four percent of amino-acid substitutions were tolerated at the dimer interface in lambda repressor.[38] In C5a,[33] interleukin-2[16] and *arc* repressor,[23] 70%, 50% and 15%, respectively, of the site-directed mutants were indistinguishable from the wild type in function or activity.

Less data is available on frequencies of up-mutations, although the first experiments—construction of second-site revertants by chemical mutagenesis *in vivo*—were done in the 1960's. Recently, the availability of long synthetic oligonucleotides and the development of methods for construction of large numbers of mutants through recombinant DNA techniques has made it feasible to generate fairly complete libraries of point mutants for any given gene. Knowles et al. showed that for triosephosphate isomerase (TIM) among a theoretical number of 4000 point mutants (the monomer of TIM is 250 amino acids long; an estimated 85% of all possible point mutants were generated), six pseudo-revertants of a 1000-fold down-mutant were obtained.[12,24] The revertants had specific activity increased by a factor of 100, 30, 6, 4, 3 and 1.5x, respectively. This complex experimental result will be represented in the model by assuming that the average increase of k_{cat} was 4-fold, and the frequency of up-mutants adjusted for constant increase of k_{cat} is about 1 per 100 mutants. These two values are consistent with the results of the β-galactosidase mutagenesis (discussed below) and will be used as parameters in further considerations.

Based on the results discussed above, we assume in the model that the frequencies of obtaining silent, down- and up-mutation are, respectively:

$$P_= = 1/2, \qquad P_{\text{down}} = 1/2, \qquad \text{and } P_{\text{up}} = 1/100 \qquad \text{Definition 1}$$

and, in addition, that the fitness function (arbitrary units) can assume discrete values of L*s, where s is a constant (step), and L = 1,2,... represents different fitness levels of the system. A protein can occupy any of these levels. From any given level L, upon the introduction of a point mutation, the protein can move to any one of two neighboring fitness level, L-1 or L+1, or stay at the level L. For consecutive fitness levels, the numerical value of the fitness function decreases/increases by a factor of 4. This assumption is based on the results of mutagenesis of the triosephosphate isomerase, discussed above. This assumption is consistent with the β-galactosidase mutagenesis results discussed below, and other multiple mutation experiments.

MULTIPLE MUTATIONS

The number of possible point mutations of a given protein, on the order of 10^3–10^4, is certainly much less than the current capacity to construct recombinant clones, which can be estimated at more than 10^{10}, and perhaps as much as 10^{20}. Thus, the ability to predict fractions of mutants in the given specific activity range (i.e., distribution of the fitness function) for multiple mutants is not only of theoretical interest, but can aid mutagenesis experiments.

TABLE 1 Effects of multiple mutations[1]

Fitness Function	Number of Mutations					
	0	I	II	III	IV	V
1000						0.0001
256					0.01	0.02
64				1	2	3
16			100	150	150	130
4		10,000	10,000	7,600	5,200	3,500
1	1,000,000	500,000	260,000	140,000	78,000	42,000
1/4		500,000	500,000	380,000	270,000	180,000
1/16			250,000	380,000	380,000	330,000
1/64				125,000	250,000	320,000
1/256					62,000	160,000
1/1024						31,000

[1] The numbers in the table give an average frequency of obtaining the particular fitness function value per 1,000,000 protein molecules mutated (see column 1) after introduction of a defined number of point mutation (indicated in roman numerals (see columns 2 thru 7)). The calculated frequencies are a result of applying Definition 1 to the presented formulation of the model.

TABLE 2 β-gal experiment vs. model

% Activity	Frequency[1] (experiment) %	Frequency (model) %
>100	1	0.3
25–100	20	4
6–25	27	18
2–6	21	33
<2	31	45

[1] Data tabulated from Ref. 14.

The model allows for a simple calculation of the distribution of the fitness function values for multiple mutants by iteration of probabilities of silent, up- and down-mutations (Table 1). The results in Table 1 can be compared with the experimental results of Dunn et al.,[4] who randomly mutagenized the alpha-complementing fragment of beta-galactosidase (90 a.a. residues). Mutagenesis was done separately in two windows, 5 a.a. each, thus five amino acids were changed at a time. A large number of mutants were randomly picked, and the activity of the complemented beta-galactosidase was measured. The cumulative data for two independently mutated windows indicate that 1% of the mutations were up-mutations, 20% of the mutants showed activity higher than 25% of the wild type activity, and 50% of the mutants had activity less than 6% of the wild-type peptide. The comparison of experimental and calculated frequencies summarized in Table 2 indicate adequate agreement. The measured specific activity of the mutants was on average reduced to 0.15 of the wild type. Assuming that each mutation is an independent event, the expected reduction of the specific activity per point mutation is therefore 0.68 $(=0.15^{1/5})$, compared to 0.67 $(=0.5*0.25+0.5*1+0.01*4)$ from the model.

Similarly, in recent experiments several simultaneous point mutations were constructed in lambda repressor[27] and *Staphylococcus* nuclease.[15] Five residues in the hydrophobic core region of the lambda repressor were simultaneously mutated in a random fashion. About 0.1% to 0.5% of the mutants had binding properties equal to the wild-type repressor. Five successive amino-acid residues within a beta-turn of *Staphylococcus* nuclease were randomly mutated. About 5% of the population of the clones scored positive for the nuclease activity. Both of these results are in a reasonable agreement with predictions based on the model (the fraction of active mutants equals (0.5),[17] i.e., 3%).

The model presented has been built using data derived from a small volume of the sequence space, namely that which corresponds to a Hamming distance of 5 or less from a small set of natural proteins or their point mutants, and to the fitness function values not higher than 3-fold, and not less than 1/1000 that of activities of natural proteins. Because of the adequate agreement of the model with experiment, it is reasonable to postulate the extension of the model onto the whole sequence space. Thus, it is further assumed that *the probabilities for decreasing, maintaining, or increasing the fitness of the polypeptide are the same at every point in the sequence space, except for the global optimum(a), and are given by Definition (1)* The fitness function is limited by physical and chemical restrictions for the binding or chemical processes (e.g., diffusion of the substrate). The assumption has far reaching consequences as to general properties of the fitness space and characteristics of walks in the fitness space (see *Implications of the model* section).

OVERLAPPING GENES

The model developed above will be applied to the evolution of single genes and overlapping genes. It was discovered in late 1970's that a single DNA fragment can encode not only one protein, but even two polypeptides.[7,36] The list of known overlapping genes is presented in Table 3. It does not present a difficulty in most cases to distinguish the evolutionarily younger gene of the two, since it is shorter, frequently lies within the older one, and encodes a protein that performs a simple function, such as binding. In the most common situation, the overlapping genes are transcribed in the same direction, but their mRNAs are translated in different reading frames. Overlapping genes present an interesting evolutionary problem of how the same DNA sequence can evolve to encode two and, in some cases, even three[10] or perhaps four[33] proteins.

In parallel overlapping genes the probability that a protein sequence change in one gene will change the sequence of the other gene is, according to calculations based on the genetic code table, approximately 66%. If one gene has evolved prior to the evolution of the second one, one half of these sequence changes, according to the model, will be deleterious to the protein encoded by the older gene. Thus only 34% + 0.5*66%, i.e., 67% of available mutations will be tolerable, and potentially beneficial, to the gene system. This value should be compared to 100% for a single gene system (there is no other gene that could be affected). The rate of evolution of the younger protein is then 67% of the rate of evolution of non-overlapping genes from random sequences. Equivalently, the probability of an up increment in f.f. value upon a point mutation will be not 1/100, but rather 1/150. Similarly, the probability that the f-f value will stay unchanged or will be reduced is 0.5*0.66 = 0.33, and 0.66, respectively. The probabilies for triple overlapping genes will be, in the same order, 1/200, 0.22, and 0.77. Thus, the relative frequencies of an up-, silent and down-mutations for single, overlapping, and triple overlapping genes (walk parameters) are: 1:50:50, 1:50:100, and 1:50:175, respectively. The significantly lower rates will manifest themselves in smaller numbers of overlapping genes.

All possible protein sequences in the sequence space occupy, according to the model, different and discrete fitness levels. To be consistent with the walk parameters for single genes, the ratio of populations of two neighboring levels is 50[1] (100 for overlapping genes, as simple calculations can show). The number of levels clearly has to be less than 230, since at this number the elements of the sequence space are exhausted ($50^{230} = 20^{300}$, 300 a.a. is the length of the average protein). Actually, the number of levels can be much less than 230, in such case the population of the maximum fitness level will be large.

---

[1]Population of level L is, according to definition (1): $N_L = N_{L-1}/100 + N_L/2 - N_{L+1}/2$, i.e., $N_L = N_{L-1}/50 - N_{L+1}$. If $N_{L+1} << N_L, N_{L-1}$, then $N_L = N_{L-1}/50$.

TABLE 3 Known overlapping genes and their properties[1]

Species	Gene 1, Function	Length (codons)	Gene 2, Function	Length (codons)	Overlap (codons)	Reference
Bacteriophage G4	Gene B replication phage assembly	120	Gene A	554	120	17
	Gene E lysis protein	96	Gene D	152	96	17
	Gene K function unknown	56	Gene A + gene C	554 152	56	17
Phage MS2	lysis protein	74	coat protein		74	35
Influenza virus	NS1 nonstructural		NS2 non-structural	70	36, 37	
Sendai virus	Protein C' non-struct.	204	Protein P structural	568	204	38
Plasmid RK2, E. coli	trfB repressor	101	incC incompatibility	364	101	39
Herpes simplex virus	mRNA B 21K polypep.		mRNA C 33K polypep.	110	40	
Phage F105 B. subtilis	ORF2 gene		immunity repressor		5	41

TABLE 3 (continued)

Species	Gene 1, Function	Length (codons)	Gene 2, Function	Length (codons)	Overlap (codons)	Reference
Phage lambda	I gene		K-gene		34	18
	xis		int		8	18
Mitochondria	URFA6L		ATPase6		8	42
Hepatitis B virus (HBV)	polymerase	830	surface antigen	387	387	28
HIVII	6 overlapping sequences, average length of 50 a.a.					27
Triple overlapping genes						
HIVII	env, tat and art proteins, triple overlap of 31 a.a.					27

[1] Data shown for those overlapping genes for which the presence of the gene product was demonstrated, but the functtion may be unknown. Data is not provided for genes which are highly homologous to those given in the Table (e.g.,HIVI, or HBV isolates from different species).

It is known from experiment[2] that the ratio of populations of evolved single and overlapping genes is on the order of 100 to 1,000. This number corresponds in the model to the ratio of the respective populations at the top fitness level. If single and overlapping genes evolved by sampling of the same pool of random sequences, but under different search parameters, followed by memorizing the most fit sequences, the non-optimized sequences cannot be further than 7 to 10 fitness levels down from the top level. Indeed, if the population of the bottom level (level 0) is normalized to 1, then the population of the next higher level is 1/50 for the single genes and 1/100 for overlapping genes, and it is, in general, $(1/50)^L$ or $(1/100)^L$, respectively, where L is the level number. The ratio of the two gene populations at any given fitness level is, therefore, 2^L. Since the ratio is known from experiment (100–1,000), the level number can be calculated, and it is 7 to 10.

This conclusion can be translated to a statement that it is sufficient to sample on the order of $50^7 = 10^{12}$ to $50^{10} = 10^{17}$ random polypeptides of about 100 a.a. in length (similar to the length of overlapping genes, Table 3) to obtain, on the average, one optimized protein with a desired simple biological activity.

In an alternative approach, the genome of a given virus can be analyzed for single genes, as well as double or triple (if applicable) overlapping sequences. The lengths and frequencies of the genes are then used as a basis for calculating the probabilities for evolving a gene from non-optimized sequences. The genome of HIVII[10] provides 9 genes encoding polypeptides with a combined length of 2,500 a.a., 6 overlapping sequences (average length of 50 a.a.), and one 31 a.a. triple overlapping sequence, which encodes portions of the *env*, *tat* and *art* proteins. Since the ratio of single to overlapping gene length is 8 to 1, we deduce the number of fitness levels to be around three. Similarly, from the ratio of the length of double and triple overlapping genes, which is equal to 10 to 1, the number of levels is again three. The results imply that a protein 30–50 a.a. in length can be selected from the pool of $50^3 = 10^5$ random polypeptides.

The estimate based on the analysis of a single viral genome yields a numerical value significantly lower than that for all protein sequence data available (10^5 versus 10^{12}–10^{17}). The factors that complicate the above analysis are: (a) the different polypeptide lengths and (b) the fact that some of the gene overlaps discussed apparently originated by an extension of an existing single gene into another pre-existing single gene (e.g., *gag* and *pol* genes of HIV). The process of gene extension can be multistep, and at each step selection pressure can be applied to assure the fitness of the protein(s) and the system. Since no attempt has been made to include selection into the analysis, the estimate obtained (10^5) is probably skewed low. In any case, the estimates are within a range that is, in principle, accessible by current molecular biology techniques.

[2]There are 12,700 entries in PIR protein database[41] as of September 1989, which number includes about 2,500 viral sequences; average length of an entry—270 a.a. The number of overlapping genes as listed in Table 3 is 20. Thus, the ratio of non-overalpping to overlapping genes ranges from 2,500/20 to 12,700/20, depending on the method of estimation.

IMPLICATIONS OF THE MODEL

There are no local maxima in the fitness space as described in the model. Indeed, the probability of the fitness function not having a higher or equal value within distance of 1 is negligibly small ($0.519 * 300$ per average length protein). It follows that any two points in the fitness space can be connected by a smooth walk (along which local optima are not encountered). In other words, any polypeptide sequence can be evolved in a monotonic (step-by-step) fashion to exhibit a desired simple biological activity. The number of steps is fairly small. Since there is a multiplicity of maxima of the same fitness value, one per 10^{12} to 10^{17} random sequences, the number of the fitness levels is small, $L = 7$ to 10. Thus, to evolve a short, 100 a.a. in length, protein from a random sequence, not more than 10 individual amino-acid replacements are sufficient to turn a random sequence into a functional protein, provided that one knows exactly which replacements to make.

Alternatively, an evolved or optimized protein, which occupies the highest fitness level allowed, can be subjected, according to the model, to a drift of its amino acid sequence without any apparent effect on its biological activity. Such a drift is indeed common for many proteins (e.g., HIV proteins, cytochrome c, etc.).

The model can be considered at best a first-level approximation to the complexities of the protein sequence space. Its formulation is extremely simple, which allows for making experimentally testable predictions. At the same time it is not rigorous, since the fitness function (or an example of it) was not explicitly defined over the sequence space, nor is a proof given that such a definition is possible. Some of the predictions or implications of the model may be counterintuitive (the smoothness of the sequence space, extremely large number of sequences that perform the same function, low number of sequence changes for evolving a random sequence, and others). At the same time, the predictions are consistent with available experimental data.

CONSTRUCTION OF RANDOMIZED GENES AND PROTEINS
OUTLINE

An alternative approach to theoretical considerations is the design of a library of random gene sequences, and then attempting the selection of a biologically active protein from the pool of random polypeptides. There are several technical obstacles in the construction of large random gene libraries. First, random DNA sequences will have stop codons distributed in intervals on the average of 64 bp. For average gene length of 1,000 bp there would be 16 stop codons, and the probability of generating a 1,000 bp open reading frame (ORF) from random sequences is about 10^{-7}. Since the maximal technically feasible size of a gene library can be estimated at 10^{10} sequences, construction of totally random sequences would not well serve the purpose of the experiment. Second, cloning of long synthetic DNA sequences is inefficient and error-prone when compared to the cloning of biological DNA. The reason is presumably chemical modification of synthetic oligonucleotides, leading to inefficient replication of synthetic DNA in the cell, in some cases resulting in sequence errors. The rate of undesired deletions or insertions is about 1 per 500–1,000 bp,[42] considerable for medium-size and long genes. Cloning inefficiencies may result in inadequate heterogeneity of the gene library. Third, a most widely used method of gene synthesis—assembly of double-stranded DNA from synthetic oligonucleotides—cannot be used (random sequences will not anneal in a controllable fashion), so alternative methods need to be developed or applied.

The method presented relies on the cloning of relatively short DNA sequences from synthetic oligonucleotides. These sequences are ORFs in both orientations of the DNA fragment, at the expense of excluding codons for two amino acids. These "building blocks" were ligated in a random fashion to form long ORFs, and cloned in a dedicated *E. coli* expression vector. A library of 10^6 randomized genes was obtained.

APPROACH

The principle of the method lies in the observation that the coding region consisting of $(N\ N\ C/T)_k$ codons (N is any nucleotide; C/T, either cytosine or thymine; G/A, either guanine or adenine) is an ORF in either orientation (in the other orientation it is $(G/A\ N\ N)_k$, also an ORF). The two types of codons encode 18 out of 20 amino acids, excluding glutamine and tryptophan, and do not include any of the three stop codons. To obtain biological DNA, short restriction fragments encoding 25 a.a. residues are cloned first by oligonucleotide-directed doublestrand break repair (bridge mutagenesis).[29,30] The procedure is extremely simple, does not require any enzymatic reactions *in vitro*, and utilizes a single-stranded synthetic DNA, approximately 100 nt in length. The detailed cloning protocol has been published elsewhere.[31] The number of the independent clones obtained was approximately 2,000. The double-stranded biological DNA fragments (building blocks) were cut out from the plasmid using *Hpa*II restriction endonuclease. The number of building blocks was 4,000 (each of 2,000 fragments can provide the coding sequence in two orientations).

The building blocks were ligated in a random fashion to a dedicated *E. coli* expression vector, so that multimers of 8 to 30 building blocks (600 to 2,500 bp) were combined to form a randomized gene. 10^6 independent clones were obtained by transformation.

ANALYSIS OF A GENE

One of the clones from the library of the randomized genes was selected for analysis of the sequence and the protein product of the gene. The sequence of the gene is shown in Figure 1. It is made of 8 building blocks, and encodes a protein of a predicted length of 213 a.a. (including 13 a.a. at the NH$_2$ terminus from the expression vector). Four 5'-end proximal fragments are in (G/A N N) orientation, the remainder four fragments—in the opposite orientation. The nucleotide composition is close to expected (25% each), namely 24%-A, 24%-C, 25%-G and 27%-T. Because of the presence of *Hpa*II restriction sites (CCGG recognition sequence) between fragments, a glycine residue (encoded by GGN) is present at every 25th residue (in addition to other positions). Thus, the sequence of the gene leaves no doubt that it was obtained synthetically.

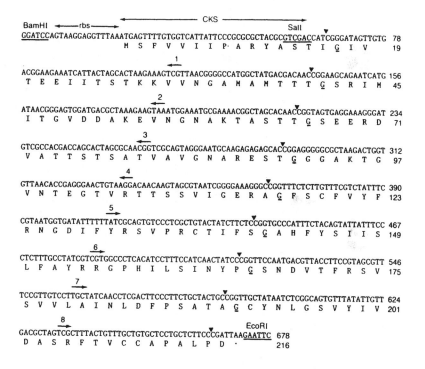

FIGURE 1 Sequence of the synthetic gene, and the predicted protein sequence. Junction points between different *Hpa*II fragments (building blocks) are indicated by triangles. Arrows indicate the relative orientations of fragments. The CKS sequence (top line) is from Miller.[33]

Predicted amino acid composition of the protein[31] is generally similar to that of an average *E. coli* protein,[8] but as expected, no glutamine and tryptophan residues are present. Both the pattern of the hydropathicity plot and of predicted secondary structure elements resemble patterns seen in natural proteins (data not shown). The protein was synthesized in *in vitro* translation system, and was identified by PAGE. The observed M_r agrees well with the predicted M_r. The protein could be expressed *in vitro* at a high level when fused to a well-expressed *E. coli*

CONCLUSIONS

The presented method is a general procedure for constructing randomized ORFs devoid of stop codons. The building block encoded 25 a.a. Nevertheless, shorter building blocks can also be used. If fragments encoding 5 a.a. sequences made out of 18 amino acids are generated, the theoretical heterogeneity ($5^{20} = 2 \times 10^6$) can be achieved experimentally, taking advantage of the improved cloning efficiency for shorter DNA fragment via bridge mutagenesis,[30] and by implementing a moderate increase of the scale of experimentation. Ligation products of such sequences will be random ORF sequences. Thus, the method permits the construction of random proteins made out of 18 a.a. (excluding glutamine and tryptophan), with the only constraint being the type of amino-acid residue at the sites of ligation.

A single synthetic oligonucleotide preparation permits construction of randomized ORFs, as described. While this may be advantageous in many cases, the method does not provide a control over the sequences being assembled. The *Fok*I method was developed[30] to address the needs of constructing defined gene sequences with a large number of degeneracies.FokI method of gene synthesis

ACKNOWLEDGMENTS

I thank Stuart Kauffman for discussions and comments, and Tom Kavanaugh and Owen McCall for reading the paper. This work was funded in part by a research grant from the Technical Advisory Board, Abbott Diagnostic Division.

REFERENCES

1. Baker, A., and G. Schatz. *Proc. Natl. Acad. Sci. USA* **84** (1987):3117–3121.
2. Dhaese, P., J. Seurinck, B. D. Smet, and M. Van Montagu. *Nuc. Acids Research* **13** (1985):5441–5455.
3. Dube, D. K., and L. A. Loeb. *Biochemistry* **28** (1989):5703–5707.
4. Dunn, I. S., R. Cowan, and P. A. Jennings. *Protein Eng.* **2** (1988):283–291.

5. Fearnley, I. M., and J. E. Walker. *EMBO J.* **5** (1986):2003–2008.
6. Fox, S. W., and K. Dose. *Molecular Evolution and the Origin of Life*, revised edition. New York: Marcel Dekker, 1977.
7. Godson, G. N., B. G. Barrell, and J. C. Fiddes. *Nature* **276** (1978):236–247.
8. Gouy, M., and C. Gautier. *Nucleic Acids Res.* **10** (1982):7055–7074.
9. Gupta, K. C., and D. W. Kingsbury. *Bioch. Bioph. Res. Comm.* **131** (1985):91–97.
10. Guyader, M., M. Emerman, P. Sonigo, F. Clavel, L. Montagnier, and M. Alizon. *Nature* **326** (1987):662–669.
11. Hamming, R.W. *Bell Syst. Tech. J.* **29** (1950):147–160.
12. Hermes, J. D., S. C. Blacklow, K. A. Gallo, A. J. Bauer, and J. R. Knowles. In *Protein Structure, Folding and Design 2*. Alan R. Liss, Inc., 1987, 257–264.
13. Homas, C. M., and C. A. Smith. *Nucleic Acids Res.* **14** (1986):4453–4469.
14. Horwitz, M. S. Z., and L. A. Loeb. *Proc. Natl. Acad. Sci. USA* **83** (1986):7405–7409.
15. Hynes, T. R., R. A. Kautz, M. A. Goodman, J.F. Gill, and R. O. Fox. *Nature* **339** (1989):73–76.
16. Ju, G., L. Collins, K. L. Kaffka, W.-H. Tsien, R. Chizzonite, R. Crowl, R. Bhatt, and P. L. Kilian. *J. Biol. Chem.* **262** (1987):5723–5731.
17. Kaiser, C. A., D. Preuss, P. Grisafi, and D. Botstein. *Science* **235** (1987):312–317.
18. Kastelein, R. A., E. Remaut, W. Fiers, and J. van Duin. *Nature* **295** (1982):35–41.
19. Kauffman, S., and S. Levin. *J. Theor. Biol.* **128** (1987):11–45.
20. Kauffman, S. A., E. D. Weinberger, and A. S. Perelson. *Theoretical Immunology*, edited by A. S. Perelson. SFI Studies in the Sciences of Complexity, Proc. Vol. II. Reading, MA: Addison-Wesley, 1988, 349–381.
21. Kim. C. W., M. L. Michaels, P. G. Markiewicz, and J. H. Miller. *J. Cell. Biochem. Suppl.* **13A** (1989):73.
22. Kleina, L. G., and J. H. Miller. *J. Cell. Biochem. Suppl.* **13A** (1989):55.
23. Knight, K. L., J. U. Bowie, A. K. Vershon, R. D. Kelley, and R. T. Sauer. *J. Biol. Chem.* **264** (1989):3639–3642.
24. Knowles, J. R. Personal communication.
25. Lamb, R. A., P. W. Choppin, R. M. Chanock, and C.-J. Lai. *Proc. Natl. Acad. Sci. USA* **77** (1980)1857–1861.
26. Lamb, R. A., and C.-J. Lai. *Cell* **21** (1980):475–485.
27. Lim, W. A., and R. T. Sauer. *Nature* **339** (1989):31–36.
28. Ma, J., and M. Ptashne. *Cell* **48** (1987):847–853.
29. Mandecki, W. *Proc. Natl. Acad. Sci. USA* **83** (1986):7177–7181.
30. Mandecki, W., and T. J. Bolling. *Gene* **68** (1988):101–108.
31. Mandecki, W. *Protein Engineering*, 1989.
32. Maynard-Smith, J. *Nature* **225** (1970):563–564.
33. Miller, R. H., S. Kaneko, C. T. Chung, R. Girones, and R. H. Purcell. *Hepatology* **9** (1989):322–327.

34. Min, K. T., M. H. Kim, and D.-S. Lee. *Nucleic Acids Res.* **16** (1988):5075–5088.
35. Mollison, K. W., W. Mandecki, E. Zuiderweg, E., Fayer, T. A., Fey, R. A., Krause, R.G. Conway, L. Miller,R. P. Edalji, M. A. Shallcross, B. Lane, J. L. Fox, J. Greer, and G. W. Carter. *Proc. Natl. Acad. Sci. USA* **86** (1989):292–296.
36. Normark, S., S. Bergstrom, T. Edlund, T. Grundstrom, F. P. Lindberg, and O. Olson. *Annu. Rev. Genet.* **17** (1983):499–525.
37. Ptitsyn, O. B., and M. V. Volkenstein. *J. Biomolec. Str. Dynam.* **4** (1986):37-156.
38. Reidhaar-Olson, J. F., and R. T. Sauer. *Science* **241** (1988):53–57.
39. Rixon, F. J., and D. J. McGeoch. *Nucleic Acids Res.* **12** (1984):2476–2487.
40. Schneider, T. D., and G. Stormo. *Nucleic Acids Res.* **17** (1989):659–674.
41. Sidman, K. E., D. G. George, and W. C. Barker. *Nucleic Acids Res.* **16** (1988):18 1871.
42. Wosnick, M. A., R. W. Barnett, A. M. Vincentini, H. Erfle, R. Elliot, M. Summer-Smith, N. Mantei, and R. W. Davies. *Gene* **60** (1987):115–127.

Marshall S. Z. Horwitz, Dipak K. Dube, and Lawrence A. Loeb
The Joseph Gottstein Memorial Cancer Research Laboratory, Department of Pathology
SM-30, University of Washington, Seattle, WA 98195

Studies in the Evolution of Biological Activity from Random Sequences

We expand upon a previous essay detailing recent advances in the selection of biologically active DNA sequences from random populations.[11] Within the framework of evolution, factors are considered that have precluded the testing of all possible sequences, purely with regard to their function as genetic regulatory elements or catalytically active RNA or protein molecules. Examples are drawn from the selection of bacterial promoters from random DNA, selection of new ribozyme activities from random RNA, and mutagenesis of enzyme domains. Efforts to derive new activities are examined, and the likelihood of future success is evaluated.

INTRODUCTION

In this essay we consider the possibility that new biological activities can be selected from cells harboring populations of random nucleotide sequences. This hypothesis assumes that evolution has not tested all permutations of nucleotides that encode a particular function, nor in the case of enzymes, has it selected, from all possible nucleotide permutations, those sequences that code for the most active enzymes. To analyze the feasibility of selecting new biological activities from random nucleotide sequences, we will first consider restrictions in the selective processes that might have occurred during prebiotic evolution. We will then review the initial attempts at selecting new active molecules from random DNA sequences and consider future possibilities for using this technique.

Let us consider a chronology of successive pressures that have been hypothesized to select for the most-fit sequences during prebiotic evolution. It has been proposed that the earliest genetic macromolecules might have contained nucleosides or similar chemical structures.[23] This notion is based on the ease by which purines and pyrimidines are formed by combustion of simple compounds that were present in the prebiotic environment. However, nitrogenous bases in this early genetic material might not have been linked by sugar phosphates; methane or ethane bridges could have constituted a primordial polynucleotide backbone.[23] When oligonucleotides did finally emerge as the genetic material, it is likely that these prototype genes were composed of RNA, with the subsequent appearance of DNA. Evidence for this includes the simplicity of the organization of present-day RNA viruses; the sequence of metabolic pathways, with ribonucleotides serving as precursors for the synthesis of deoxyribonucleotides; the nearly universal use of RNA primers for the initiation of DNA replication; and the demonstration that the splicing and joining of ribonucleotides as well as the net elongation of an RNA primer can be catalyzed by an RNA enzyme.[2] Assuming four different bases, and an average length of 800 nucleotides,[28] a hypothetical gene could be selected from a reservoir of 4^{800} possible permutations. One early determinant for selection would be the ability to form a stable double-stranded structure, something that would be favored by a high content of GC base pairs. It is of interest that damage to DNA by a variety of environmental agents occurs preferentially at guanosines and involves destabilization of the glycosylic bond with the formation of an abasic site[17] and that DNA polymerases preferentially incorporate deoxyadenosine opposite abasic sites.[29] If the nucleic acid containing the abasic site codes for the preferential incorporation of deoxyadenosine triphosphate by DNA polymerase, this could have provided a mechanism for the gradual conversion of GC-enriched sequences to those with all four nucleotides present.

Important parameters for the survival of a ribonucleotide sequence or a set of overlapping sequences (a "quasispecies") would likely have included the rate of template-directed nucleotide assembly and the fidelity of this process.[4] The substitution of deoxynucleotides for ribonucleotides in the genome would have resulted in increased stability and a decreased frequency in the production of branched molecules during replication. Transcription and translation are likely to be later events in prebiotic evolution and could logically coincide with the formation of cells. Selection with respect to protein structure and enhanced catalytic activity is likely to be a relatively late event that occurred upon the emergence of complex cellular structures. As a result of these sequential selective processes, many nucleotide substitutions that would be advantageous for catalysis may no longer be able to be tested by the substitutions involving only one or a few nucleotides.

REPLACEMENT OF GENETIC REGULATORY ELEMENTS BY RANDOM SEQUENCES

In order to select biologically active DNA sequences from populations of random DNA, one desires a short segment of DNA that both encodes a selectable biological activity and can tolerate nucleotide substitution at high frequency. The promoter region of procaryotic genes (for review see Horwitz and Loeb[12]) is particularly suitable for these studies, since this region controls the expression of genes that could be selectable and is also composed of a consensus sequence that, by definition, can tolerate substitutions. In our earliest studies we have substituted 16 or 19 basepairs

	-35						-10						
G	1	6	54	3	6	9	3	1	11	7	6	3	**G**
C	3	5	5	24	42	14	15	2	14	8	24	1	**C**
A	0	5	1	46	13	18	1	67	19	37	29	3	**A**
T	86	74	30	17	26	45	53	2	28	20	13	65	**T**

	Spacer					Start Site			
15	**16**	**17**	**18**	**19**	**5**	**6**	**7**	**8**	**9**
2	20	100	27	5	3	12	31	1	1

FIGURE 1 Consensus matrix of synthetic *E. coli*-promoter elements. The number of times each base occurs in each position and the number of those elements separated by the given base spacing is indicated. These are the combined results of the 170 promoters presented in Horwitz and Loeb,[9] Horwitz and Loeb,[10] and Oliphant and Struhl.[21] Start site spacing refers to the distance between the -10 site and +1, and these data are available only from the first two of the above references.

centered at the -35 promoter element of the plasmid pBR322 tetracycline resistance gene with chemically and enzymatically synthesized random DNA sequences to produce a population of plasmid DNA in which each molecule contains a unique sequence.[9,10] This population of molecules was transfected into *E. coli* to produce random sequence libraries from which growth selection with the antibiotic identified those sequences, from among the 3×10^{11} possible, with promoter activity. Oliphant and Struhl[21] have subsequently conducted similar studies employing histidine prototrophy as a selection. A promoter consensus sequence can be derived from the 170 sequences that we and the latter workers have produced (Figure 1). Even though this consensus sequence is derived from analyzing some populations of promoter substitutions that were biased in base composition, the spectrum of functional promoters is remarkably similar to the native consensus sequence.

The consensus sequence and spacing of the random promoters differs from the consensus derived from 263 natural promoters and traditional mutations thereof[7] only in its preference for a T, in place of an A, at the most downstream position along the -35 element. Among our more interesting conclusions is the observation that the very strongest promoters departed somewhat from the consensus and revealed unexpected homology in some ordinarily nonconserved spacer sequences. A similar random sequence analysis has since been conducted on bacteriophage T7 promoters.[30]

SELECTION OF NOVEL RIBOZYMES FROM RANDOM RNA SEQUENCES

A unique substrate for the design of novel catalysts from random sequence populations is RNA, since it has both phenotypic and genotypic characteristics. That is, RNA has both the potential for catalytic activity as well as encoding the information necessary for its reproduction. Joyce and coworkers have developed a method for the selective amplification of individual RNA molecules exhibiting specific catalytic functions.[13] They have identified an RNA enzyme with the capability of cleaving DNA via a transesterification reaction.[27] The substrate DNA is hybridized to a conserved segment common to a population of random RNA molecules (Figure 2). If, among the random RNA molecules, are those that form a structure with catalytic activity, they will cleave the DNA and ligate a portion of the DNA to the 3' ribonucleotide terminus. The covalent RNA-DNA hybrid can then be amplified by alternate cycles of cDNA synthesis with reverse transcriptase (using a primer complementary to the ligated region) and transcription with a phage RNA polymerase (taking advantage of a promoter sequence encoded in the original randomized RNA population). Note that the catalytic steps also serve to facilitate the ultimate characterization of the active species. This highly imaginative approach can be used for the design of new enzymes that recognize and alter the sequence of any RNA or DNA molecule.

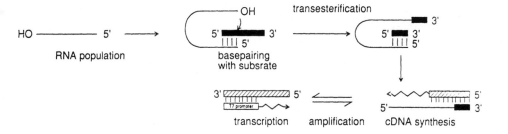

FIGURE 2   Selection of catalytically active RNA molecules from randomized sequences. In this example[27] selection is for DNA cleavage via a transesterification reaction to produce a hybrid RNA-DNA molecule that may be amplified by alternate rounds of reverse and forward transcription.

USE OF RANDOM SEQUENCES FOR DELINEATING FUNCTIONAL GROUPS IN PROTEINS

Of all the amino acids in an enzyme, only a few directly interact with the substrate. Perhaps the bulk of the residues function only to maintain the globular stability of the protein (for example, Matsumura et al.[20]). Based on the results of experiments using site-specific mutagenesis with diverse enzymes, it can be inferred that the number of amino-acid residues that alter the rate or specificity of catalysis is limited; a reasonable estimate for the number of nucleotides that specify the active site would be 60. Thus, potential mutations that effect catalysis in a typical enzyme would be 4^{60}, a number that is small compared to the total number of possible permutations (4^{800}). Therefore, a large number of potential nucleotide sequences that are no longer easily evolved from the cellular repertoire might contain information to code for enzymes with greater catalytic potency or with different specificities than those that are naturally present.

In our initial studies[3] we have remodeled the active site of the gene encoding β-lactamase, the enzyme that cleaves the β-lactam ring of penicillin antibiotics. The replacement sequence ($Phe^{66}XXXSer^{70}XXLys^{73}$) preserves the codon for the active serine 70, but also contains 15 base pairs of random sequences coding for 3.2×10^6 amino-acid substitutions. From a population of *E. coli*-harboring plasmids with these substitutions, we identified 7 new active-site mutants that render bacteria resistant to β-lactam antibiotics. Subsequently, Oliphant and Struhl[22] have performed similar experiments in which they have substituted the region from Arg^{61} to Cys^{77} with sequences that were not entirely random but contained on the average 20% base mutations at each position (about 3 mutant codons per molecule). For the most part, the results of these studies are similar and are summarized in

Figure 3. From this similarity of results the following conclusions can be made: (1) Mutations in this region do indeed alter substrate specificities for the differently substituted β-lactam antibiotics. (2) Ser[70] is required, as no substitution of this residue was possible in our study, nor was it obtained in the other study. (3) Most substitutions are temperature sensitive. This surprising result suggests that evolutionary selection for thermal stability may be almost as important as catalytic activity. (4) A great many of substitutions are tolerable.

One major difference in these two studies is the frequency of mutations at Thr[71]. The high frequency of substitutions that we observed is supported by the studies of Schultz and Richards[31] who demonstrated that 14 of the 20 amino acids can functionally substitute this position. Oliphant and Struhl's failure to detect more than one change here may indicate unevenness in the design of their experimental strategy.

One structural feature of protein molecules that has proven particularly amenable to replacement by random sequences are leader sequences. These amino terminus sequences consist of a limited number of amino acids that facilitate the export

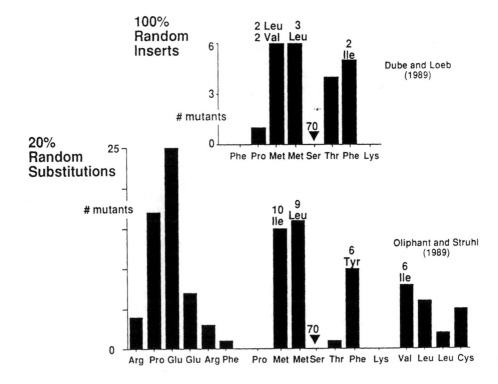

FIGURE 3   Histogram of amino-acid substitutions in the active site of β-lactamase.

of proteins in both procaryotic and eucaryotic cells. A comparison of 150 known leader sequences indicates a diversity of functional sequences as well as an element of commonality, the most apparent being a requirement for a stretch of at least seven hydrophobic amino acid residues.[32] Kaiser et al.[14] replaced leader sequences of the gene that codes for the secreted yeast invertase gene with random fragments derived from human genomic DNA. After transfecting the reconstructed gene into yeast, they observed that about 20% of the transformants contained sequences that function in the export of invertase. Moreover, there were sequences that lacked the diagnostic hallmark of a leader sequence, stretches of hydrophobic amino-acid residues. In related experiments, Baker and Schatz[1] have shown that random digests of the *E. coli* genome can functionally replace at similar high frequency the signal sequence involved in the mitochondrial targeting of yeast cytochrome c oxidase subunit IV.

In still other examples, Ma and Ptashne[18] fused a library of *E. coli* genomic DNA fragments to the coding sequence of the DNA-binding portion of the yeast GAL4 gene. After transfection into yeast, they obtained a series of new activating sequences that showed no obvious homology to one another or with the activating regions of the GAL4 gene. The only common feature among the polypeptides that are attached to the DNA-binding domain of GAL4 is that they are grossly acidic. Recently, Gill, Sadowski, and Ptashne[6] have identified multiple amino-acid substitutions scattered over a 65-residue region of GAL4 that result in enhanced rates of transcriptional activity compared to the wild type. This is not the only example of randomly directed mutations resulting in increased function. Hermes et al.[8] have shown that residue replacements throughout the length of triose-phosphate isomerase are capable of enhancing catalytic activity in a mutant enzyme, and we have found that random sequence replacement of the thymidine binding site of the herpes virus thymidine kinase identifies new mutants that exhibit increased kinase activities *in vivo*. Purification and kinetic studies on these putative high-activity kinases are required to substantiate these findings and are in progress.

FUTURE PROSPECTS

The technique of selecting biologically active DNA sequences from large random populations provides a new method for identifying nucleotide sequences with unique functions. Perhaps its greatest utility is in the engineering of new proteins with novel activities. Since many enzymes are far from catalytically perfect (i.e., $k_{cat}/K_m < 10^8 - 10^9$ mol^{-1} second(s)$^{-1}$, the diffusion rate limit of catalytic velocity), it may be possible to select many new sequences that are more active in catalysis than those that are found naturally. Alternatively, remodeling of substrate binding sites may broaden the substrate specificity of enzymes. Particularly promising is the recent discovery and engineering of catalytic antibodies (reviewed by Powell and Hansen[24]). Random sequence mutation of such molecules may prove productive.

The ultimate challenge of genetic engineering with random sequences will be the construction of completely new enzymes. Several thermodynamic and theoretical analyses of polypeptide folding have come to the conclusion that many random sequences may well have stable conformations.[15,16,25,26]. Fox and Dose[5] have shown that thermally induced polymerization of amino acids can produce a few molecules with catalytic activity, including that of proteolysis. It may thus be possible to select a novel enzyme from a large population of random polypeptides. Toward this end, Mandecki[19] has produced a library of 10^6 random proteins with an average molecular weight of 30 kDa, capable of being expressed in *E. coli*.

ACKNOWLEDGMENTS

This work was supported by National Institutes of Health grants R35CA39903 (to Lawrence A. Loeb) and GM07266 (to the University of Washington Medical Scientist Training Program).

REFERENCES

1. Baker, A., and G. Schatz. "Sequences From a Prokaryotic Genome or the Mouse Dihydrofolate Reductase Gene can Restore the Import of a Truncated Precursor Protein into Yeast Mitochondria." *Proc. Natl. Acad. Sci. U.S.A.* **84** (1987):3117–3121.
2. Been, M. D., and T. R. Cech. "RNA as an RNA Polymerase: Net Elongation of an RNA Primer Catalyzed by the *Tetrahymena* Ribozyme." *Science* **239** (1988):1412–1416.
3. Dube, D. K., and L. A. Loeb. "Mutants Generated by the Insertion of Random Oligonucleotides into the Active Site of the β-Lactamase Gene." *Biochem.* **28** (1989):5703–5707.
4. Eigen, M., and P. Schuster. "The Hypercycle, A Principle of Natural Self-Organization." *Naturwissenschaften* **64** (1977):541–565.
5. Fox, S. W., and K. Dose. *Molecular Evolution and the Origin of Life*, revised edition. New York, NY: Marcel Dekker, 1977.
6. Gill, G., I. Sadowski, and M. Ptashne. "Mutations that Increase the Activity of a Transcriptional Activator in Yeast and Mammalian Cells." *Proc. Natl. Acad. Sci. U.S.A.* **87** (1990):2127–2131.
7. Harley, C. B., and R. P. Reynolds. "Analysis of *E. coli* Promoter Sequences." *Nucl. Acids Res.* **5** (1987):2343–2361.

8. Hermes, J. D., S. C. Blacklow, and J. R. Knowles. "Searching Sequence Space by Definably Random Mutagenesis: Improving the Catalytic Potency of an Enzyme." *Proc. Natl. Acad. Sci. U.S.A.* **87** (1990):696–700.

9. Horwitz, M. S. Z., and L. A. Loeb. "Promoters Selected From Random DNA Sequences." *Proc. Natl. Acad. Sci. U.S.A.* **83** (1986):7405–7409.

10. Horwitz, M. S. Z., and L. A. Loeb. "DNA Sequences of Random Origin as Probes of *E. coli* Promoter Architecture." *J. Biol. Chem.* **263** (1988):14724–14731.

11. Horwitz, M. S. Z., D. K. Dube, and L. A. Loeb. "Selection of New Biological Activities from Random Nucleotide Sequences: Evolutionary and Practical Considerations." *Genome* **31** (1989):112–117.

12. Horwitz, M. S. Z., and L. A. Loeb. Structure-Function Relationships in *Eschericiacoli* Promoter DNA." *Prog. Nucl. Acids Res. Mol. Biol.* **38** (1990): 137–164.

13. Joyce, G. F. "Amplification, Mutation and Selection of Catalytic RNA." *Gene* **82** (1989):83–87.

14. Kaiser, C. A., D. Preuss, P. Grisafi, and D. Botstein. "Many Random Sequences Functionally Replace the Secretion Signal Sequence of Yeast Invertase." *Science* **235** (1987):312–317.

15. Kauffman, S. A. "Autocatalytic Sets of Proteins." *J. Theoret. Biol.* **119** (1986):1–24.

16. Lau, K. F., and K. A. Dill. "Theory for Protein Mutability and Biogenesis." *Proc. Natl. Acad. Sci. U.S.A.* **87** (1990):638–642.

17. Loeb, L. A. "Apurinic Sites as Mutagenic Intermediates." *Cell* **40** (1985)483–484.

18. Ma, J., and M. Ptashne. "A New class of Yeast Transcriptional Activators." *Cell* **51** (1987)113–119.

19. Mandecki, W. "A Method for Construction of Long Randomized Open Reading Frames and Polypeptides." *Prot. Engin.* **3** (1990):221–226.

20. Matsumura, M., W. J. Becktel, and B. W. Matthews. "Hydrophobic Stabilization in T4 Lysozyme Determined Directly by Multiple Substitutions of Ile 3." *Nature* **334** (1988):406–410.

21. Oliphant, A. R., and K. Struhl. "Defining the Consensus Sequences of *E. coli* Promoter Elements by Random Selection." *Nucl. Acids. Res.* **16** (1988):7673–7683.

22. Oliphant, A. R., and K. Struhl. "An Efficient Method for Generating Proteins with Altered Enzymatic Properties Application to β-lactamase." *Proc. Natl. Acad. Sci. USA* **86** (1989):9094–9098.

23. Orgel, L. E. "RNA Catalysis and the Origins of Life." *J. Theoret. Biol.* **123** (1986):127–133.

24. Powell, M. J., and D. E. Hansen. "Catalytic Antibodies–A New Direction in Enzyme Design." *Prot. Eng.* **3** (1989):69–75.

25. Ptitsyn, O. B. "Protein as an Edited Statistical Copolymer (translation)." *Molekol. Biol. (U.S.S.R.)* **18** (1984):574–590.

26. Ptitsyn, O. B., and M. V. Volkenstein. "Protein Structures and Neutral Theory of Evolution." *J. Biomolec. Str. Dynam.* **4** (1986):137–156.

27. Robertson, D. L., and G. F. Joyce. "Selection *in vivo* of an RNA Enzyme that Specifically Cleaves Single-Stranded DNA." *Nature* **344** (1990):467–468.

28. Savageau, M. A. "Proteins of *E. coli* Come in Sizes that are Multiples of 14 kDa: Domain Concepts and Evolutionary Implications." *Proc. Natl. Acad. Sci. U.S.A.* **83** (1986):1198–1202.

29. Schaaper, R. M., and L. A. Loeb. "Depurination Causes Mutation in SOS-Induced Cells." *Proc. Natl. Acad. Sci. U.S.A.* **78** (1981):1773–1777.

30. Schneider, T. D., and G. D. Stormo. "Excess Information at B Bacteriophage T7 Genomic Promoters Detected by a Random Cloning Technique." *Nucl. Acids. Res.* **17** (1989):659–674.

31. Schultz, S., and J. H. Richards. "Site-Saturation Studies of β-Lactamase: Production and Characterization of Mutant β-Lactamase with all Possible Amino Acid Substitutions at Residue 71." *Proc. Natl. Acad. Sci. U.S.A.* **83** (1986):1588–1592.

32. von Heijne, G. "Signal Sequences: The Limits of Variation." *J. Mol. Biol.* **184** (1985):99–105.

Debra L. Robertson and Gerald F. Joyce
Department of Molecular Biology, Research Institute of Scripps Clinic, 10666 N. Torrey
Pines Rd., La Jolla, California 92037, USA

The Catalytic Potential of RNA

INTRODUCTION

Until recently the central dogma of biology held that all the catalytic functions of
unicellular and multicellular organisms are carried out by specific proteins. This
vast array of proteins is produced according to specific instructions encoded in
the nucleic acids (DNA, RNA) of the organism. In 1982, T. Cech and coworkers
showed that an RNA molecule from a ciliated protozoan could act like a protein
and catalyze the biochemical reactions necessary for its own maturation.[36] The
discovery of catalytic RNA molecules (ribozymes) has led to a re-examination of
biological catalysis and of the role of RNA in cellular metabolism.

Most of the genes in eukaryotic organisms are interrupted by stretches of
noncoding sequences referred to as "intervening sequences" or "introns." RNA
molecules are copied (transcribed) from the DNA in a precursor form that contains
both the genetic information (exons) necessary for the synthesis of a particular
protein and the introns. The precursor RNA molecules are then converted to their
mature form by a splicing process that results in removal of the introns. Most RNA
molecules require the presence of specialized proteins in order to undergo splicing
and subsequent maturation. However, certain catalytic RNAs have been shown to
self-splice *in vitro* using a mechanism that is independent of a protein catalyst.

In the past, our understanding of biological catalysis had been limited to protein enzymes. Ribozymes provide an alternative means by which biochemical reactions can be carried out. Because RNA is an informational molecule as a well as a catalytic molecule, it can be manipulated in the laboratory at the level of both genotype and phenotype. Thus, it provides a new and often simpler approach to the study of structure-function relationships involving complex macromolecules. Further studies may lead to the recognition of RNA molecules with novel catalytic properties.

Since the initial discovery of catalytic RNA, several types of ribozymes have been found. Here we review the structure and catalytic properties of the known RNA enzymes. Also, we discuss efforts to identify new RNA enzymes in biological systems and to develop new RNA enzymes in the laboratory.

THE DISCOVERY OF CATALYTIC RNA

Ribozymes were originally described while studying the removal of introns from precursor ribosomal RNA (rRNA) molecules of the ciliated protozoan *Tetrahymena pigmentosa*.[36] T. Cech and coworkers found that the splicing of a particular rRNA intron does not require any of the nuclear proteins required by other RNAs for splicing and instead requires only a guanosine co-factor and Mg^{2+}. At the same time, S. Altman, N. Pace, and coworkers were studying the RNA-processing enzyme of *E. coli* known as RNase P. This enzyme is a ribonucleoprotein (containing both RNA and protein) that specifically cleaves the precursor form of tRNA to generate a mature tRNA. It was found that the RNA component of RNase P is responsible for its catalytic activity.[19] Subsequent investigation led to the discovery of additional self-splicing RNAs in a variety of biological systems including yeast mitochondria,[46] *Neurospora* mitochondria,[17] T4 phage[5,12] and the slime mold *Physarum polycephalum*.[41]

RNA self-splicing is an intramolecular event whereby the intron acts upon itself to alter its own covalent structure. The accurate excision of the intron and the concomitant ligation of the exons is accomplished by a single catalytic activity that resides within the intron. The requirements for self-splicing have been defined by *in vitro* studies. It has yet to be shown that a ribozyme can undergo self-splicing within the cellular environment of its parent organism.

Although a self-splicing RNA resembles a protein enzyme in that it is a macromolecule that accelerates the basal rate of a specific chemical reaction, it was not considered to be a true enzyme until it was shown that it could operate on a substrate other than itself.[28,50] A modified form of the self-splicing *Tetrahymena* ribozyme was shown to act on an external oligonucleotide substrate whose sequence resembles that of the precursor RNA. Using a short oligopyrimidine substrate, which mimics the sequence at the 3' end of the 5' exon, the *Tetrahymena* ribozyme is able to carry out two intermolecular reactions that are mechanistically equivalent to self-splicing. During these *"trans*-splicing" reactions the intron remains unchanged and thus is able to turn over and catalyze additional *trans*-splicing events.[26,56]

Ribozymes, like their protein counterparts, share certain general catalytic strategies. The RNA molecule must fold into an appropriate three-dimensional structure in order to express its catalytic activity. If the molecule is incubated at high temperature[54] or in the presence of denaturing agents such as urea or formalin,[1] it loses its three-dimensional structure and undergoes a corresponding loss of catalytic activity. Specific mutations in structurally important regions of the molecule also result in a loss of activity.[2,9] Although the three-dimensional structure has not been determined for any ribozyme, it has been possible to model two-dimensional (secondary) structure based on phylogenetic comparison of closely related sequences from various organisms. This secondary structure model can in turn be used to predict the tertiary structure of the molecule.[34]

A second general catalytic strategy shared by all known ribozymes is that they exhibit a high level of substrate specificity. In the case of the self-splicing RNA from *Tetrahymena*, there is a sequence of pyrimidines at the 3' end of the 5' exon that is recognized by a sequence of purines within the catalytic apparatus of the intron.[13] Mutations that disrupt this interaction destroy the self-splicing activity of the RNA.[3,49]

Analysis of catalytic RNAs from many organisms has led to the recognition of four major categories of ribozymes. Members of each category share common structural features and operate by a similar chemical mechanism. The characteristics that differentiate each of the four categories are summarized in Table 1 and are described in detail below.

TABLE 1 Structural and Catalytic Properties of Known RNA Enzymes

	catalytic activity	chemical mechanism	reaction products	cofactors
group I introns	self-splicing	phosphoester transfer	5'-P & 2'(3')-OH	Mg^{2+}, G_{OH}
δ group II introns	self-splicing	phosphoester transfer	5'-P & 2'(3')-OH	Mg^{2+}
δ RNase P	tRNA processing	hydrolysis	5'-P & 2'(3')-OH	Mg^{2+}
δ hammerheads	self-cleavage	hydrolysis	2'(3')-P & 5'-OH	Mg^{2+}

GROUP I INTRONS

In the group I intron addition to the *Tetrahymena* ribozyme there are approximately seventy intervening sequences that have been categorized as group I introns based on the presence of conserved sequences and secondary structural features.[13,39,40] Of these seventy only eight have been shown to be self-splicing *in vitro*. Self-splicing occurs via a two-step mechanism involving successive transesterification reactions (Figure 1(A)). Transesterification results in exchange of phosphodiester bonds at specific sites in the RNA. There is no net change in the number of ester linkages and therefore no need for an external energy source such as ATP. In the first transesterification reaction the 5' splice site is cleaved and guanosine is added to the 5' end of the intron. The 2'(3') hydroxyl located at the end of the released 5' exon then acts as a nucleophile to attack the 3' splice site. The 3' splice site is cleaved and the exons are joined together to yield a mature RNA molecule while releasing the full-length intron.[53,54] Accuracy of the self-splicing reaction is insured by: 1) an "internal guide sequence" located near the 5' end of the intron that binds the sequence located at the 3' end of the 5' exon; and 2) a "G binding site" within the catalytic center of the intron that binds the guanosine residue located at the 3' end of the intron.[24]

The secondary structure of group I introns is characterized by nine base-paired regions designated P1–P9 (Figure 2).[8] These regions can fold into a consensus structure that is presumed to be responsible for the molecule's catalytic activity. The most highly conserved regions constitute a core structure that is thought to contain the active site of the ribozyme (shown in bold in Figure 2).[11] The core resembles a pseudoknot structure: it is an RNA hairpin whose 3' end loops around and forms a second helix within the hairpin.

Deletion analysis has been used to define those regions of the intron that are essential for self-splicing activity.[29,32,45] The core structure, the internal guide sequence, and the 3' terminal G residue are all required, while structural elements that extend from these regions can for the most part be deleted (Figure 2). Folding of the active site of the molecule depends on the formation of secondary and tertiary structural interactions involving these essential regions.

In addition to self-splicing, the *Tetrahymena* ribozyme has been shown to catalyze a number of reactions that are fundamentally equivalent to self-splicing. For example, it can mediate cleavage/ligation reactions involving two dinucleotides or a mononucleotide and a trinucleotide, so long as those substrates conform to the sequence specificity of the self-splicing reaction.[33] The *Tetrahymena* ribozyme can extend pentacytidylic acid (C_5) when supplied with either GC or GU to form C_5C_n or C_5U_n, respectively.[4] It can also act as a sequence-specific endonuclease to perform a cleavage reaction at a phosphodiester bond following a sequence of pyrimidines.[55]

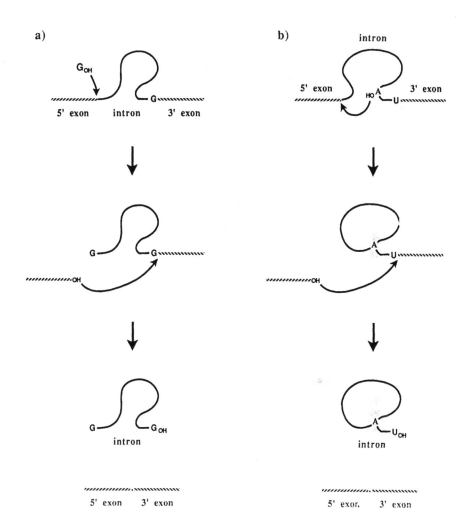

FIGURE 1 Self-splicing of group I and group II introns. (A) Group I self-splicing occurs via two transesterification events. First, guanosine acts as a nucleophile to cleave the 5′ splice site. The released 5′ exon then acts as a nucleophile to cleave the 3′ splice site, resulting in release of the intron and ligation of the exons. (B) Self-splicing of group II introns is initated by the 2′-hydroxyl of an adenosine residue within the intron. The intron is released as a "lariat" structure.

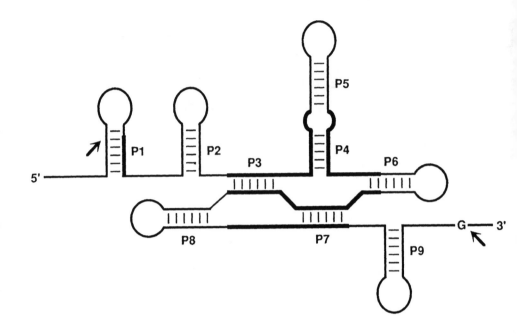

FIGURE 2 Secondary structure of group I introns labeled according to the standard nomenclature.[8] Arrows designate the splice sites. The internal guide sequence (located in element P1) and the conserved core are shown in bold.

GROUP II INTRONS

There is a second group of introns found in chloroplasts and yeast mitochondria that share a set of conserved sequences and secondary structure distinct from that of group I introns. These are referred to as group II introns.[40] Certain group II introns have been shown to be self-splicing *in vitro*.[43] Like the self-splicing group I introns they require Mg^{2+}, but unlike group I introns do not require guanosine or any other nucleoside cofactor. Group II self-splicing also occurs via two successive transesterification reactions; the first involving cleavage at the 5′ splice site and the second involving cleavage at the 3′ splice site to produce the ligated exons (Figure 1(B)). Unlike group I introns, which make use of a guanosine 2′(3′) hydroxyl to initiate the first transesterification reaction, group II introns utilize the 2′ hydroxyl of a nucleotide within the RNA chain to initiate self-splicing. The excised intron is left as a branched (lariat) structure that contains an internal 2′,5′ phosphodiester bond (Figure 1(B)).[43,46,47]

The splicing mechanism in group II introns is closely related to that which occurs during the splicing of nuclear mRNAs.[48] Like group II introns, nuclear mRNA molecules form a "lariat" intermediate. However, nuclear mRNAs do not self-splice and instead rely on a complex apparatus known as the "spliceosome" to carry out splicing. The intron of a nuclear pre-mRNA facilitates splicing by providing three conserved regions that demarcate the two splice sites and the lariat position. Deletions of large portions of the intron outside of these conserved regions do not affect mRNA splicing.[42] This is in contrast to group I and group II introns whose catalytic activity is dependent on the overall structure of the intron.

RNASE P

The activity of catalytic RNA molecules is not restricted to self-splicing. Ribonuclease P (RNase P) is a ribonucleoprotein that accurately removes the 5' leader sequence from a newly formed pre-tRNA to yield a mature tRNA.[22,35] RNase P can recognize other RNA substrates, including certain RNA viruses, that contain structural features that resemble the characteristic cloverleaf structure of tRNA.[18,21,38]

The structure of RNase P RNA has been studied by comparing two different versions of the enzyme from the eubacteria *E. coli* and *B. subtilis*. Despite the low sequence similarity that exists between the RNA from these two organisms, a common primary and secondary structure has been identified.[27] Deletion analysis of the RNase P RNA from *E. coli* has further defined the critical structural features.[51]

The catalytic activity of RNase P is dependent on the structure of the RNA and the presence of Mg^{2+}. Like self-splicing group I introns, RNase P generates reaction products that have 5'-phosphate and 2'(3')-hydroxyl termini.[19] Unlike that of group I or group II introns, the mechanism of action does not involve nucleophilic attack by a nucleoside 2'(3')-hydroxyl. Modification of the terminal 3' hydroxyl of RNase P RNA does not alter its catalytic activity. It has been suggested that the RNA binds Mg^{2+} which coordinates a free hydroxyl ion that acts as the nucleophile to carry out the hydrolysis reaction.[20,37]

HAMMERHEADS

The "hammerhead" subclass of ribozymes describes a small group of plant viroids, virusoids and linear satellite RNAs that share a common secondary structure.[16] The original discovery involved the virusoid of satellite tobacco ringspot virus which was found to undergo a specific cleavage reaction during replication.[44] The self-cleavage reaction is used to produce linear monomeric forms of the virusoid from a polymeric form that is generated by rolling-circle replication.[6] Like the reactions catalyzed by group I and II introns and RNase P, the hammerhead self-cleavage reaction requires a divalent cation and a well-defined secondary and tertiary structure. Structural analysis of these RNAs by Symons and coworkers established a "hammerhead" model which suggests that the catalytic domain consists of three short RNA helices supporting a highly conserved region of 13 nucleotides.[16] The avocado sunblotch viroid (ASBV), lucern transient streak virusoid (VLTSV), and newt satellite RNA are other examples of catalytic RNAs that fit into the hammerhead structural model.[15,16,25] Unlike the reactions carried out by group I and II introns and RNase P, the hammerhead reaction results in products that have 5′-hydroxyl and 2′(3′)-phosphate termini.

There are two examples of small catalytic RNAs that, like the hammerheads, undergo a self-cleavage reaction to produce 5′-hydroxyl and 2′(3′)-phosphate termini at the cleavage site. Analysis of their secondary structure shows a catalytic domain that does not conform to the hammerhead model. The first example is the hepatitis delta agent, a small satellite RNA of hepatitis B virus. When the delta agent superinfects cells containing the hepatitis B virus, it causes an especially severe form of the disease. The hepatitis delta agent undergoes self-cleavage via the same mechanism as the hammerheads.[52] Although the catalytic domain of the delta agent has not been established, it is clear from analysis of its sequence and possible secondary structure that it is not a hammerhead.

The negative strand tobacco ringspot virus ((-)TRSV) is another plant pathogen that undergoes a self-cleavage reaction to generate products that have 5′-hydroxyl and 2′(3′)-phosphate termini.[23] The catalytic domain of (-)TRSV includes two well-defined regions of 50 nucleotides and 14 nucleotides that distinguishes it from both the hammerheads and the hepatitis delta agent. It would not be surprising if additional examples of self-cleaving RNAs come to be recognized among RNA viruses and virus-associated RNAs.

NEW RIBOZYMES

The identification of four categories of ribozymes has caused investigators to focus attention on cellular reactions that involve RNA in an effort to identify novel ribozymes. One likely site is the spliceosome. The spliceosome is a structure that is involved in the removal of introns from precursor RNA molecules. The spliceosome contains the precursor RNA molecule and four small nuclear ribonucleoproteins, designated U1, U2, U5 and U4/6. Each of these ribonucleoproteins contains a small nuclear RNA. The involvement of these small RNAs in RNA splicing has led to speculation concerning their possible catalytic role in the splicing process.[7]

The replication of poliovirus may provide another system where a catalytic RNA is involved. The genomic RNA of poliovirus (and other picornaviruses) encodes a small polypeptide (VPg) that becomes covalently bound to its 5′ terminus during replication. Attachment of VPg to the viral RNA requires only Mg^{2+}, VPg polypeptide and the viral RNA itself.[10] Attachment occurs via a transesterification reaction involving nucleophilic attack by a tyrosine residue of the VPg at a phosphodiester bond near the 5' terminus of the viral RNA. This reaction is reminiscent of nucleophilic attack by guanosine at the 5' splice site that occurs during the first step of splicing in group I introns. Whether VPg attachment is catalyzed by the viral RNA alone or is co-catalyzed by VPg and the viral RNA remains to be determined.

RIBOZYME ENGINEERING

Given the high level of interest in catalytic RNA, but given the fact that only a few RNA enzymes are known to exist in biology, considerable attention has been directed toward the construction of novel RNA enzymes in the laboratory. The general approach has been to modify an existing ribozyme in order to alter its substrate specificity or modify some aspect of its catalytic behavior. Been and Cech changed the internal guide sequence of the *Tetrahymena* ribozyme from GGAGGG to GAAAAG, and as a result changed its substrate preference from oligo(C) to oligo(U).[3] More recently, Doudna and Szostak were able to externalize the guide sequence and generalize its template properties so that the ribozyme could act on substrates containing any of the four nucleotide bases.[14] One of the goals of this line of investigation is to generalize the substrate requirements to such an extent that the ribozyme can act on any external template, including a copy of itself, and thus act as a general purpose replicase. Such a molecule, when provided with appropriate energy-rich starting materials, would replicate autonomously.

In our laboratory we are taking a somewhat different approach. We have developed *in vitro* techniques for the rapid selection, amplification and mutation of catalytic RNA.[31] A heterogeneous population of RNA molecules is tested for their ability to perform a specific catalytic task. Reactive molecules are selected and amplified subject to mutational error to produce a new generation of variant RNAs. By iterating this process of selection, amplification and mutation we are attempting to evolve novel RNA enzymes in the laboratory.[30] In the strictest sense, applied molecular evolution of this type should not be referred to as "engineering." We can only guide the evolving population by progressively altering the selection constraints, allowing the system to discover acceptable solutions for itself.

A third approach, one that has not yet been applied to RNA enzymes, is to construct the catalyst *de novo*. At the present time we do not know enough about the tertiary structure of RNA enzymes to recognize common structural motifs that could serve as the basis for a rational approach to ribozyme design. Both group I introns and RNase P RNA contain a pseudoknot structure within their catalytic center. However, it is not known whether the pseudoknot is an essential feature of these molecules, let alone how one should go about attaching additional structural elements to a pseudoknot core. If catalysis is not an extraordinarily rare property of RNA molecules, then it may be possible to select for catalytic RNAs beginning with a population of random oligonucleotides or beginning with a heterogeneous population of RNAs that share a common structural foundation. If RNA catalysis is a special property that requires a long period of evolutionary development, then existing ribozymes are likely to provide the best starting point for exploring the catalytic potential of RNA.

REFERENCES

1. Bass, B. L., and T. R. Cech. "Biological Catalysis by RNA." *Ann. Rev. Biochem.* **55** (1986):599–629.
2. Been, M. D., and T. R. Cech. "Sites of Circularization of the Tetrahymena rRNA IVS are Determined by Sequence and Influenced by Position and Secondary Structure." *Nucl. Acids Res.* **13** (1985):8389–8409.
3. Been, M. D., and T. R. Cech. "One Binding Site Determines Sequence Specificity of Tetrahymena Pre-rRNA Self-Splicing, Trans-Splicing and RNA Enzyme Activity." *Cell* **47** (1986):207–216.
4. Been, M. D., and T. R. Cech. "RNA as an RNA Polymerase: Net Elongation of an RNA Primer Catalyzed by the Tetrahymena Ribozyme." *Science* **239** (1988):1412–1416.
5. Belfort, M., J. Pederson-Lane, D. West, K. Ehrenman, and G. Maley. "Processing of the Intron-Containing Thymidylate Synthetase (td) Gene of Phage T4 is at the RNA Level." *Cell* **41** (1985):375–382.
6. Branch, A. D., H. D. Robertson, and E. Dickson. "Longer-Than-Unit-Length Viroid Minus Strands are Present in RNA from Infected Plants." *Proc. Natl. Acad. Sci. USA* **78** (1981):6381–6385.
7. Brow, D. A., and C. Guthrie. "Splicing a Spliceosomal RNA." *Nature* **337** (1989):14–15.
8. Burke, J. M., M. Belfort, T. R. Cech, R. W. Davies, R. J. Schweyen, D. A. Shub, J. W. Szostak, and H. F. Tabak. "Structural Conventions for Group I Introns." *Nucl. Acids Res.* **159** (1987):7212–7221.
9. Burke, J. M. "Molecular Genetics of Group I Introns: RNA Structure and Protein Factors for Splicing—A Review." *Gene* **73** (1988):273–293.
10. Cech, T. R. "Ribozymes and Their Medical Implications." *J. Am. Med. Assoc.* **260** (1988):3030–3034.
11. Cech, T. R. "Conserved Sequences and Structures of Group I Introns: Building an Active Site for RNA Catalysis—A Review." *Gene* **73** (1988):259–271.
12. Chu, F. K., G. F. Maley, F. Maley, and M. Belfort. "Intervening Sequence in the Thymidylate Synthetase Gene of Bacteriophage T4." *Proc. Natl. Acad. Sci. USA* **81** (1984):3049–3053.

13. Davies, R. W., R. B. Waring, J. A. Ray, T. A. Brown, and C. Scazzocchio. "Making Ends Meet: A Model for RNA Splicing in Fungal Mitochondria." *Nature* 300 (1982):719–724.
14. Doudna, J. A., and J. W. Szostak. "RNA-Catalyzed Synthesis of Complementary-Strand RNA." *Nature* 339 (1989):519–522.
15. Epstein, L. M., and J. G. Gall. "Self-Cleaving Transcripts of Satellite DNA from the Newt." *Cell* 48 (1987):535–543.
16. Foster, A. C., and R. H. Symons. "Self-Cleavage of Plus and Minus RNAs of a Virusoid and a Structural Model for the Active Sites." *Cell* 49 (1987):211–220.
17. Garriga, G., and A. M. Lambowitz. "RNA Splicing in Neurospora Mitochondria: Self-Splicing of a Mitochondrial Intron in vitro." *Cell* 39 (1984):631–641.
18. Green, C. J., B. S. Vold, M. D. Morch, R. L. Joshi, and A. Haenni. "Ionic Conditions For the Cleavage of the tRNA-Like RNA of the Turnip Yellow Mosaic Virus by the Catalytic RNA Component of RNase P." *J. Biol. Chem.* 263 (1988):11617–11620.
19. Guerrier-Takada, C., K. Gardiner, T. Marsh, N. Pace, and S. Altman. "The RNA Moiety of Ribonuclease P is the Catalytic Subunit of the Enzyme." *Cell* 35 (1983):849–857.
20. Guerrier-Takada, C., K. Haydock, L. Allen, and S. Altman. "Metal Ion Requirements and other Aspects of the Reaction Catalyzed by M1 RNA, the RNA of Ribonuclease P from Escherichia coli." *Biochemistry* 25 (1986):1509–1514.
21. Guerrier-Takada, C., A. van Belkum, C. W. A. Pleij, and S. Altman. "Novel Reactions of RNase P with a Transfer RNA-like Structure in Turnip Yellow Mosaic Virus RNA." *Cell* 53(2) (1988):267–272.
22. Guthrie, C., and R. Atchison. "Biochemical Characterization of RNase P: A tRNA Processing Activity with Proteins and a RNA." In *Transfer RNA: Biological Aspects*, edited by D. Soll, J. N. Abelson, and P. R. Schimmel. New York: Cold Spring Harbor Laboratory, 1980, 83–97.
23. Hampel, A., and R. Tritz. "RNA Catalytic Properties of the Minimum (-)sTRSV Sequence." *Biochemistry* 28 (1989):4929–4933.
24. Hanna, M. F., M. Green, R. Bartel, D. P. and J. W. Szostak. "The Guanosine Binding Site of the Tetrahymena Ribozyme." *Nature* 342 (1989):391–395.
25. Hutchins, C. J., P. D. Rathjen, A. C. Forester, and R. H. Symons. "Self Cleavage of Plus and Minus RNA Transcripts of Avocado Sunblotched Viroid." *Nucl. Acids Res.* 14 (1986):3627–3640.
26. Inoue, T., F. X. Sullivan, and T. R. Cech. "Intermolecular Exon Ligation of the rRNA Precursor of Tetrahymena: Oligonucleotides Can Function as 5 Exons." *Cell* 43 (1985):431–437.
27. James, B. D., G. J. Olsen, J. Lin, and N. R. Pace. "The Secondary Structure of Ribonuclease P RNA, the Catalytic Element of a Ribonucleoprotein Enzyme." *Cell* 52 (1988):19–26.

28. Jencks, W. P. Catalysis. In *Chemistry and Enzymology*. New York: McGraw-Hill, 1969.
29. Joyce, G. F., and T. Inoue. "Structure of the Catalytic Core of the Tetrahymena Ribozyme as Indicated by Reactive Abbreviated Forms of the Molecule." *Nucl. Acids Res.* **15** (1987):9825–9840.
30. Joyce, G. F. "Building the RNA World: Evolution of Catalytic RNA in the Laboratory." In *Molecular Biology of RNA*, edited by T. R. Cech. UCLA Symposium on Molecular and Cellular Biology, vol. 94. New York: Alan R. Liss, 1988, 361–371.
31. Joyce, G. F. "Amplification, Mutation, and Selection of Catalytic RNA." *Gene* **82** (1989):85–87.
32. Joyce, G. F., G. van der Horst, and T. Inoue. "Catalytic Activity is Retained in the Tetrahymena Group I Intron Despite Removal of the Large Extension of Element P5." *Nucl. Acids Res.* **17** (1989):7879–7889.
33. Kay P. S., and T. Inoue. "Catalysis of Splicing-Related Reactions Between Dinucleotides Catalyzed by a Ribozyme." *Nature* **327** (1987):343–346.
34. Kim, S-H., and T. R. Cech. "Three Dimensional Model of the Active Site of the Self-Splicing rRNA Precursor of Tetrahymena." *Proc. Natl. Acad. Sci.* **84** (1987):8788–8792.
35. Kole, R., M. F. Baer, B. C. Stark, and S. Altman. "*E. coli* RNAase P has a Required RNA Component." *Cell* **19** (1980):881–887.
36. Kruger, K., P. J. Grabowski, A. J. Zaug, J. Sands, D. E. Gottschling, and T. R. Cech. "Self-Splicing RNA: Autoexcision and Autocyclization of Ribosomal RNA Intervening Sequence of Tetrahymena." *Cell* **31** (1982):147–157.
37. Marsh, T. L., and N. R. Pace. "Ribonuclease P Catalysis Differs from Ribosomal RNA Self-Splicing." *Science* **229** 79–81.
38. McClain, W. H., C. Guerrier-Takada, and S. Altman. "Model Substrates for an RNA Enzyme." *Science* **238** (1987):527–530.
39. Michel, F., A. Jacquier, and B. Dujon. "Comparasion of Fungal Mitochondrial Introns Reveals Extensive Homologies in RNA Secondary Structures." *Biochemie* **64** (1982):867–881.
40. Michel, F., and B. Dujon. "Conservation of RNA Secondary Structures in Two Intron Families Including Mitochondrial-, Chloroplast-, and Nuclear-Encoded." *EMBO J.* **2** (1983):33–38.
41. Muscarella, D. E. and V. M. Vogt. "A Mobile Group I Intron in the Nuclear rDNA of Physarum polycephalum." *Cell* **56** (1989):443–454.
42. Padgett, R. A., P. J. Grabowski, M. M. Konarska, S. Seiler, and P. A. Sharp. "Splicing of Messenger RNA Precursors." *Ann Rev. Biochem.* **55** (1986):1119–1150.
43. Peebles, C. L., P. S. Perlman, K. L. Mecklenburg, M. L. Petrillo, J. H. Tabor, K. A. Jarrell, and H. L. Cheng. "A Self-Splicing RNA Excises as an Intron Lariat." *Cell* **44** (1986):213–223.
44. Prody, G. A., J. T. Bakos, J. M. Buzayan, I. R. Schneider, and G. Bruening. "Autolytic Processing of Dimeric Plant Virus Satellite RNA." *Science* **231** (1986):1577–1580.

45. Szostak, J. W. "Enzymatic Activity of the Conserved Core of a Group I Self-Splicing Intron." *Nature* **322** (1986):83–86.
46. Van der Horst, G., and H. F. Tabak. "Self-Splicing of Yeast Mitochondria Ribosomal and Messenger RNA Precursors." *Cell* **40** (1985):759–766.
47. Van der Veen, R., A. C. Arnberg, G. van der Horst, L. Bonen, H.F. Tabak, and L.A. Grivell. "Excised Group II Introns in Yeast Mitochondria are Lariats and Can Be Formed by Self-Splicing in vitro." *Cell* **44** (1986):225–234.
48. Waring, R. B., C. Scazzocchio, T. A. Brown, and R. W. Davies. "Close Relationship Between Certain Nuclear and Mitochondria Introns: Implications for the Mechanism of RNA Splicing." *J. Mol. Biol.* **167** (1983):595–605.
49. Waring, R. B., P. Towner, S. J. Minter, and R. W. Davies. "Splice Site Selection by a Self-Splicing RNA of Tetrahymena." *Nature* **321** (1986):133–139.
50. Walsh, C. *Enzymatic Reaction Mechanisms.* San Francisco: Freeman, 1979.
51. Waugh, D. S., L. J. Green, and N. R. Pace. "The Design and Catalytic Properties of a Simplified Ribonuclease P RNA." *Science* **244** (1989):1569–1571.
52. Wu, H., and M. M. C. Lai. "Reversible Cleavage and Ligation of Hepatitis Delta Virus RNA." *Science* **243** (1989):652–654.
53. Zaug, A. J., and T. R. Cech. "The Intervening Sequence Excised from the Ribosomal RNA Precursor of Tetrahymena Contains a 5 Terminal Guanosine not Encoded by the DNA." *Nucl. Acids Res.* **10** (1982):2823–2838.
54. Zaug, A. J., P. J. Grabowski, and T. R. Cech. "Autocatalytic Cyclization of an Excised Intervening Sequence is a Cleavage-Ligation Reaction." *Nature* **310** (1983):578–583.
55. Zaug, A. J., M. D. Been, and T. R. Cech. "The Tetrahymena Ribozyme Acts like an RNA Restriction Endonuclease." *Nature* **324** (1986):42–433.
56. Zaug, A. J., and T. R. Cech. "The Intervening Sequence RNA of Tetrahymena is an Enzyme." *Science* **231** (1986):470–475.

Origin of Life Models

P. Schuster
Institut für Theoretische Chemie der Universität Wien, Währingerstraße 17, A-1090 Wien, Austria

Dynamics of Autocatalytic Reaction Networks

A classification of autocatalytic reaction networks is presented. They are properly subdivided into two major classes: quasi-linear and nonlinear reaction networks. Quasi-linear networks have globally stable stationary states and no complex dynamical phenomena except travelling waves in excitable media are observed.

The nonlinear networks are more interesting from the physicist's point of view: under suitable conditions they show very rich dynamics. Bistabilities, oscillations, deterministic chaos, travelling waves, stable dissipative structures, and complex spatio-temporal patterns may occur.

The introduction of mutation into error-free nonlinear autocatalytic networks commonly simplifies their dynamics. One example of deterministic chaos which changes first into oscillations and then into a stable stationary state is reported.

AUTOCATALYSIS AND COMPLEX DYNAMIC PHENOMENA

A network of processes is considered an *autocatalytic reaction network* if elements (I_k) are produced by a copying mechanism. The elements are thus capable of self-reproduction. They are predominantly viewed as polynucleotide molecules—RNA or DNA—but they may be equally well virions, cells, or even organisms. The notion of a *replicator*—originally invented by Richard Dawkins[5] and frequently used in biology for *"an entity that passes on its structure largely intact in successive replications"*[39]—is a useful characterization of the most relevant property of the elements considered here. We shall adopt the notion of replicators whenever we discuss issues which are not restricted to molecules.

Autocatalytic reaction networks are properly subdivided into two classes: *quasi-linear* and essentially *nonlinear* networks. The properties of quasi-linear replication networks were described in another contribution to this volume.[33] In short the most important results are:

- The kinetic differential equations of quasi-linear systems can be transformed into a linear ODE by a simple transformation of the time-axis.
- Replication with finite accuracy results in a distribution of replicators which commonly consists of one dominating—*fittest*—type and a *cloud of mutants* formed by its close relatives. *Relatedness* refers here to the frequency of mutation.
- The ODE of quasi-linear systems has one stable fixed point. The stationary distribution of replicators corresponding to this fixed point—the time-independent mutant cloud—is called the *quasispecies*. In the limit of error-free replication the quasispecies consists of a single type. This is the fittest replicator which—in molecular terms—is commonly denoted as *master sequence*.
- The structure of the quasispecies depends in characteristic manner on the frequency of mutation. The mutant cloud becomes larger with increasing error rates and the relative amount of master sequence in the population decreases. At error rates above a critical value called the *error threshold*, mutations become so frequent that the master sequence looses its dominating role and differences in fitness are no longer reflected by the long-time distribution of replicators.

The remarkable fact about error propagation in quasi-linear ensembles of replicating molecules is this existence of a sharply defined error threshold which sets a limit to stable information transfer from generation to generation. In case the error rate exceeds a critical replication accuracy, the population starts migrating randomly in sequence space and there exists no stable distribution of replicating elements.

Nonlinear replication networks—in contrast to the quasi-linear systems—show an enormous variety in their dynamics which extends from monotonous approach towards a stable stationary state to bistability, complicated oscillations, chaotic dynamics, and spontaneous spatial pattern formation. The role of mutations in nonlinear replication systems is ambivalent: in most cases replication errors simplify the dynamics of the network but there are also special cases in which the opposite effect is observed.

GROWTH FUNCTIONS OF AUTOCATALYTIC REPLICATION NETWORKS

Forgetting about mutation in the first place the act of replication can be expressed by a formal reaction equation of the type

$$(\mathbf{A}) + \mathbf{I}_k \xrightarrow{\mathcal{F}_k} 2\mathbf{I}_k \; ; \quad k = 1, 2, \ldots, n \, .$$

The collection of materials from which the replicators are built is denoted symbolically by $\mathbf{A}$. We are dealing here exclusively with conditions under which the available amount of $\mathbf{A}$ is constant at every instant and hence it does not enter as a variable into the kinetic differential equations. Accordingly we write $\mathbf{A}$ in parentheses.

The replication efficiency of a particular replicator $\mathbf{I}_k$ is generally expressed as a function of concentrations $c_k = [\mathbf{I}_k]$: $\mathcal{F}_k(\mathbf{c})$ with $\mathbf{c} = (c_1, c_2, \ldots, c_n)$. Particle numbers expressed by $N_k = c_k \cdot V$ serve this purpose equally well—V is here the volume of the system. The functions $\mathcal{F}_k(\mathbf{c})$ are understood as generalized reaction-rate parameters since they comprise a whole series of uncatalyzed and catalyzed replication processes. Application of mass action kinetics yields a polynomial expansion:

$$\mathcal{F}_k(\mathbf{c}) \; = \; A_k + \sum_{j=1}^{n} A_{kj} c_j + \sum_{i=1}^{n} \sum_{j=1}^{n} A_{kj} c_j c_i \ldots , \quad k = 1, 2, \ldots, n \, . \quad (1a)$$

We are dealing with a quasi-linear replication network if only the constant terms of the expansions are considered. Truncation of the expansion after the linear terms in the functions $\mathcal{F}_k(\mathbf{c})$ yields the simplest essentially nonlinear replication networks. Here we shall omit the constant terms and study mainly networks whose replication functions $\mathcal{F}_k(\mathbf{c})$ are linear. These systems will be characterized as second-order autocatalytic reaction networks since the individual catalytic processes involve two replicators which fulfil different tasks: $\mathbf{I}_k$ acts as the template—as it does in the first-order process—and $\mathbf{I}_j$ acts as the catalyst.

It is often useful to define a *growth function* which describes the instantaneous rate of production of the replicator $\mathbf{I}_k$. In our notation this rate is expressed by $c_k \cdot \mathcal{F}_k(\mathbf{c})$.

If only terms of the same order in the expansion (1a)—first, second, third ...—are considered, the dynamics of the reaction network is—as shown in previous studies[8]—independent of the total concentration $c = \sum_{k=1}^{n} c_k$ up to a transformation of the time axis. Thus we can use normalized or relative concentrations

x_k as variables without loosing generality. They are easily expressed in terms of conventional concentrations c_k or particle numbers N_k:

$$x_k = \frac{c_k}{\sum_{i=1}^{n} c_i} = \frac{N_k}{\sum_{i=1}^{n} N_i} \quad \text{and} \quad \sum_{k=1}^{n} x_k = 1 . \tag{2}$$

Concentrations—or particle numbers—are non-negative quantities by definition and thus the domain of physically meaningful state vectors $\mathbf{x}$ is restricted to the unit simplex

$$\mathbf{S}_n \doteq \left\{ \mathbf{x} = (x_1, x_2, \ldots, x_n) \in \mathbf{R}^n : x_k \geq 0, \sum_{k=1}^{n} x_k = 1 \right\} .$$

Due to the normalization condition, the n-dimensional replication ensemble has only $n-1$ degrees of freedom.

In the normalized concentrations that we use throughout this paper, the replication function has the general form:

$$\mathcal{F}_k(\mathbf{x}) = A_k + \sum_{j=1}^{n} A_{kj} x_j + \sum_{i=1}^{n} \sum_{j=1}^{n} A_{kj} x_j x_i \ldots , \quad k = 1, 2, \ldots, n . \tag{1b}$$

This equation differs from (1a) in a minor, but nevertheless, important detail: relative concentrations are numbers and hence all rate constants A_k, A_{kj}, ..., have the dimension of a reciprocal time, $[t^{-1}]$.

MUTATION

In order to distinguish error-free replication and mutation we introduce a mutation matrix Q whose elements determine the frequencies of mutation: $Q_{k\ell}$ represents the fraction of replications with $\mathbf{I}_\ell$ as template which yield $\mathbf{I}_k$ as error copy. Since replication has to be necessarily either correct or erroneous, Q is a stochastic matrix:

$$\sum_{k=1}^{n} Q_{k\ell} = 1 \quad \forall \quad k = 1, 2, \ldots, n . \tag{3}$$

Error-free copying and mutations are understood as parallel reactions:

$$(\mathbf{A}) + \mathbf{I}_k \xrightarrow{Q_{kk}\mathcal{F}_\ell} 2\mathbf{I}_K ; \quad k = 1, 2, \ldots, n \tag{4a}$$

$$(\mathbf{A}) + \mathbf{I}_\ell \xrightarrow{Q_{k\ell}\mathcal{F}_k} \mathbf{I}_k + \mathbf{I}_\ell ; \quad k, \ell = 1, 2, \ldots, n . \tag{4b}$$

It is important to realize that no rules of mutation have been laid down yet. Depending on whether we restrict errors to point mutations as in a previous chapter[33] or whether we include also insertions and deletions, the details in the structure of the mutation matrix Q are entirely different.

The use of model assumptions for mutation, however, will be inevitable in all specific applications because otherwise the large number of mutation parameters represented by the $n \times (n-1)$ independent entries of the mutation matrix Q is prohibitive. In case of the *uniform error-rate model*[33] we have for example

$$Q_{k\ell} \;=\; q^{\nu}\left(\frac{1-q}{q}\right)^{d_{k\ell}}$$

Only two parameters, the error rate per digit (q) and the Hamming distance ($d_{k\ell}$) between the two sequences of chain length ν, $\mathbf{I}_k$ and $\mathbf{I}_\ell$, are required to derive all mutation rates $Q_{k\ell}$.

SELECTION CONSTRAINT AND KINETIC DIFFERENTIAL EQUATIONS

In addition to the network of autocatalytic processes, the details of the experimental setup or the environmental conditions have to be known in order to formulate the kinetic differential equations. The setup, in essence, has to fulfil two criteria: it has to provide an open system to keep the processes away from thermodynamic equilibrium and it should introduce a selection constraint which makes the mathematical analysis sufficiently simple. Here we model a constraint which meets the experimental conditions in an *evolution reactor* as sketched in Figure 1.

An unspecific dilution flow $\Phi(t)$ controls the total concentration of replicating molecules, $c(t) = \sum_{i=1}^{n} c_i(t) \approx c_0 = const$:

$$\Phi(t) \;=\; \sum_{i=1}^{n} \mathcal{F}_i(\mathbf{x})\, x_i \; . \tag{5}$$

Here, *unspecific* means that the probability to be diluted out of the reactor within the next time interval is the same for each molecule. Thus the loss in normalized concentration of a given replicator $\mathbf{I}_k$ is $x_k \cdot \Phi(t)$ per time unit.

The flux $\Phi(t)$ is easily recognized as the mean rate of reproduction in the ensemble and thus Eq. (5) has a straightforward interpretation when we consider momentary population dynamics[7]: every element which replicates faster than the average will increase in frequency and the concentration of every element which is less efficient than the average will decrease. This principle of differential selection holds no matter how complex the replication process is in detail.

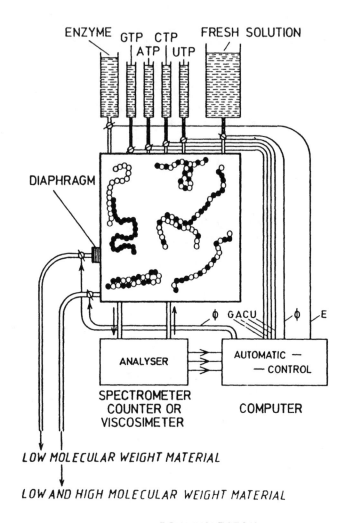

ENZYME GTP CTP FRESH SOLUTION
|ATP |UTP

DIAPHRAGM

φ GACU   φ   E

ANALYSER

AUTOMATIC —
— CONTROL

SPECTROMETER
COUNTER OR
VISCOSIMETER

COMPUTER

LOW MOLECULAR WEIGHT MATERIAL

LOW AND HIGH MOLECULAR WEIGHT MATERIAL

●........ ○●○○●●○ = POLYNUCLEOTIDE

FIGURE 1 The evolution reactor. A kind of dialysis reactor which consists of a reaction
vessel with walls which are impermeable to RNA molecules the reactor provides a flow
which keeps a reaction mixture of replicating polynucleotides, e.g., RNA molecules
$(\mathbf{I}_k; k = 1, 2, \ldots, n; [\mathbf{I}_k] = c_k)$, away from thermodynamic equilibrium. Transport
of energy rich material from the stock solutions into the reactor is adjusted in such a
way that the concentrations of the nucleoside triphosphates **GTP**, **ATP**, **CTP** and
UTP are constant in the reactor. A specific RNA replicase, commonly an enzyme
of the RNA bacteriophage $Q\beta$, is added in order to provide a medium suitable for
replication. Degradation products are removed steadily. The total concentration of RNA,
$c = c_1 + c_2 + \ldots + c_n$, is controlled by a dilution flux $\Phi(t)$. For example, a selection
constraint may be chosen which keeps the concentration c constant. Regulation of Φ
requires internal control which is achieved by automatic analysis of the solution in the
reactor and data processing by a computer which adjusts the flows in the various inlet
and outlet valves. The selection constraint of constant concentration c facilitates the
mathematical analysis of the kinetic differential equations considerably. It is commonly

The kinetic differential equations account for production by correct replication and mutation as well as for loss by the dilution flow:

$$\frac{dx_k}{dt} = \sum_{\ell=1}^{n} Q_{k\ell} \mathcal{F}_\ell(\mathbf{x}) \, x_\ell \; - \; x_k \Phi(t)$$

$$= \sum_{\ell=1}^{n} (Q_{k\ell} - x_k) \, \mathcal{F}_\ell(\mathbf{x}) \, x_\ell \; ; \quad k = 1, 2, \ldots, n \; . \tag{6}$$

The mutation matrix Q becomes an $(n \times n)$-dimensional unit matrix in the limit of vanishing replication errors: $Q \to \Im d$. Then the kinetic differential equation is of *replicator type*:

$$\frac{dx_k}{dt} = x_k \left(\mathcal{F}_k(\mathbf{x}) - \Phi \right) = x_k \left(\mathcal{F}_k(\mathbf{x}) - \sum_{j=1}^{n} \mathcal{F}_j(\mathbf{x}) \, x_j \right) \; ; \quad k = 1, 2, \ldots, n \; . \tag{7}$$

The term replicator equation was coined for these mutation-free replication systems since the same class of equations was found to be important in very different applications ranging from the theory of molecular evolution to dynamic models of games in sociobiology.[28]

QUASI-LINEAR NETWORKS

The elements of quasi-linear replication networks replicate independently of each other. The rate of replication depends exclusively on the replicator acting as template for reproduction and the expansion in Eqs. (1a) and (1b) is thus restricted to the constant term. For error-free replication and mutation leading to production of I_k, we find the rate constants $Q_{kk} A_k$ and $Q_{k\ell} A_\ell$, respectively. If reproduction is carried out in an evolution reactor (Figure 1), the selection constraint acts by the dilution flux

$$\Phi(t) = \sum_{i=1}^{n} A_i \, x_i(t) \; . \tag{5a}$$

It is obtained from Eq. (1b) by truncation after the constant term and making use of Eq. (3).

The kinetic differential equations are then of the form

$$\frac{dx_k}{dt} = \sum_{\ell=1}^{n} Q_{k\ell} A_\ell x_\ell - x_k \Phi = \sum_{\ell=1}^{n} (Q_{k\ell} - x_k) A_\ell x_\ell \; ; \quad k = 1, 2, \ldots, n \; . \tag{8}$$

The solutions and their discussions are found in the literature.[11,12,33]

The studies of deterministic kinetic differential equations were complemented by investigations of stochastic models for quasi-linear reaction networks.[6,21,22,32]

In addition more detailed stochastic models were developed in order to study the role of simple but realistic genotype–phenotype relations leading to highly *rugged* fitness landscapes.[14,15]

NONLINEAR NETWORKS

In order to be able to make full use of normalized concentrations and to derive a classification of nonlinear autocatalytic reaction networks from the kinetic differential equations, it is advisable to consider exclusively second-order reactions. The corresponding reaction equations for error-free replication and mutation are:

$$(\mathbf{A}) + \mathbf{I}_k + \mathbf{I}_j \xrightarrow{Q_{kk}A_{kj}} 2\mathbf{I}_k + \mathbf{I}_j \ ; \quad k = 1, 2, \dots, n \ . \tag{9a}$$

$$(\mathbf{A}) + \mathbf{I}_\ell + \mathbf{I}_j \xrightarrow{Q_{k\ell}A_{\ell j}} \mathbf{I}_k + \mathbf{I}_\ell + \mathbf{I}_j \ ; \quad j, k, \ell = 1, 2, \dots, n \ . \tag{9b}$$

In principle every member of this autocatalytic reaction network may act as the template—$\mathbf{I}_k$ in Eq. (9a) or $\mathbf{I}_\ell$ in Eq. (9b), respectively—and as the catalyst—$\mathbf{I}_j$ in both reaction Eqs. (9a) and (9b)—in second-order replication reactions. Catalytic action distinguishes therefore first- and second-order networks. For short we shall denote second-order autocatalytic reaction networks here as *catalytic networks*. In the networks to be studied, and also in real systems, most of the entries in the matrix of replication rate constants $A \doteq \{A_{ij}; \ i, j = 1, 2, \dots, n\}$ will be zero. For large values of n and typical situations, A is a sparse matrix.

The unspecific flux is derived from Eq. (5) whereby we make again use of the properties of the mutation matrix Q as given in Eq. (3):

$$\Phi(t) = \sum_{j=1}^{n} \sum_{i=1}^{n} A_{ij} \, x_i(t) x_j(t) \ . \tag{5b}$$

The kinetic equations for the catalytic networks can be written now in general form:

$$\frac{dx_k}{dt} = \sum_{j=1}^{n} \sum_{\ell=1}^{n} Q_{k\ell} A_{\ell j} x_\ell x_j - x_k \Phi$$

$$= \sum_{\ell=1}^{n} \left\{ (Q_{k\ell} - x_k) \sum_{j=1}^{n} A_{\ell j} x_\ell x_j \right\} \ ; \quad k = 1, 2, \dots, n \ . \tag{10}$$

Equation (10) is very hard to study directly: it is neither a replicator Eq. (7) nor can it be transformed into a linear ODE. A two-step strategy, however, turned out to be very useful: in the first step we shall analyze the error-free system and later on in the second step we shall introduce mutation by means of perturbation theory.

In the limit of vanishing mutation rates the kinetic differential equations are of replicator type:

$$\frac{dx_k}{dt} = x_k \left(\sum_{j=1}^{n} \left\{ A_{kj} x_j - \sum_{\ell=1}^{n} A_{\ell j} x_\ell x_j \right\} \right) \ ; \quad k = 1, 2, \dots, n \ . \tag{11}$$

Some properties of replicator equations like the question of *permanence* or *exclusion* have been studied in detail.

- A network is *permanent* if none of its members is lost in the limit $t \to \infty$.
- *Exclusion* takes place if the concentration of at least one replicator vanishes in the long time limit.

For a comprehensive survey of the properties of replicator equations and a presentation of many examples, see the monograph by Hofbauer and Sigmund.[20] It is important to mention a difference in notation: Hofbauer and Sigmund define the order of the replicator equation according to the degree of the polynomial in $\mathcal{F}_k(\mathbf{x})$, whereas we refer—in agreement with previous definitions[8]—to the degree of the polynomial in the growth function $x_k \cdot \mathcal{F}_k(\mathbf{x})$ which in the case of mass action kinetics is identical with the molecularity of the reactions in the network (9). Thus Sigmund and Hofbauer's order equals ours minus one. Another argument in favor of our choice might lie in the fact that the first-order replicator equation can indeed be transformed into a linear differential equation.[11,12]

Three special cases of the second-order replicator Eq. (10) are of particular interest:

1. In the multi-dimensional Schlögl model[26] the matrix of replication rate constants is diagonal: $A_{ij} = \delta_{ij} \cdot F_i$ where we make use of Kronecker's δ:

$$\delta_{ij} = \left\{ \begin{array}{ll} 1 & \text{if } i = j; \\ 0 & \text{if } i \neq j. \end{array} \right.$$

The ODE of the multi-dimensional Schlögl model can be transformed into a generalized gradient of Shahshahani type[25] and hence oscillations of concentrations or deterministic chaos can be excluded. Gradient systems are characterized by the existence of a potential function $V(\mathbf{x})$ which allows to write the differential equation in the form

$$\frac{dx_k}{dt} = \frac{\partial V\mathbf{x}}{\partial x_k} ; \qquad k = 1, 2, \ldots, n .$$

From calculus follows that the Jacobian of a gradient system is symmetric and therefore almost all trajectories converge monotonously towards asymptotically stable fixed points. The n-dimensional system has $2^n - 1$ fixed points, n of them are asymptotically stable and coincide with the corners of the concentration simplex.[29,30] The multidimensional Schlögl model represents an illustrative example of simultaneously stable stationary states. We are dealing with n coexisting point attractors, each one having its own basin of attraction whose size is determined by the rate constant F_k.

2. Fisher's selection equation[13,20] is another special case with simple dynamics which is of fundamental importance in population genetics. The laws of molecular genetics imply that the matrix of replication constants A is symmetric: $A_{ij} = A_{ji}$.

The selection equation is also a Shahshahani gradient and again oscillations as well as other types of complicated dynamics can be excluded.

3. In the hypercycle model[9] non-zero entries of the replication matrix A are restricted to cyclically permuted off-diagonal coupling terms: $A_{ij} = \delta_{i+1,j} \cdot C_j$ with i, j mod n.

Hypercycle equations lead to permanent dynamical systems.[18] They have globally stable limit cycles for $n \geq 5$. So far no strange attractor was found. Approximate analytical solution curves were computed.[23,24] For large values of n these curves are reminiscent of concentration waves travelling along a closed loop,

$$I_1 \rightarrow I_2 \rightarrow \ldots \rightarrow I_{n-1} \rightarrow I_n \rightarrow I_1 \rightarrow \ldots \, ,$$

on the boundary of the simplex S_n. In the limit of large n only three neighboring elements are present in appreciable amounts. The concentrations of all other $n-3$ replicators are very small but non-zero.

The general replicator equation has a very rich dynamics. For $n = 3$ and appropriate choice of parameters it sustains marginally stable oscillations as does the conventional Lotka-Volterra equation. A full classification of the various phase portraits was presented recently.[34] In higher dimensions, $n \geq 4$, stable limit cycles,[17] Feigenbaum-type cascades of period doublings as well as deterministic chaos were reported.[27,31]

Replicator equations are special cases of ecological equations

$$\frac{dx_k}{d} = x_k \, \mathcal{G}_k(\mathbf{x}) \; ; \quad k = 1, 2, \ldots, n \, , \tag{12}$$

where $\mathcal{G}_k(\mathbf{x})$ is some polynomial in $x_1, \ldots, x_n$. It is worth mentioning that second-order replicator equations of dimension n and Lotka-Volterra equations of dimension $n-1$,

$$\frac{dx_k}{dt} = x_k \left(\alpha_k + \sum_{j=1}^{n-1} \beta_{kj} x_j \right) \; ; \quad k = 1, 2, \ldots n-1 \, , \tag{13}$$

are equivalent[19]; they can be interconverted by means of a nonlinear transformation of variables. In certain cases it is useful to work in the replicator representation of the ODE (11) because its dynamics is confined to the unit simplex S_n and this may facilitate qualitative analysis considerably.

CLASSIFICATION OF ERROR-FREE SECOND-ORDER NETWORKS

In order to classify the dynamics of error-free catalytic networks with general second-order growth rates, we first remove an ambiguity: Equation (11) is invariant to additive constants in the columns of the replication matrix A. Two replicator equations with A and

$$A' = \left\{ A_{ij} + k_j; \; i,j = 1,2,\ldots,n \right\}$$

have identical solution curves, trajectories and phase portraits. In order to remove this arbitrariness we choose a particular *normal form* of Eq. (11) in which the replication matrix B has only zero diagonal elements $(k_j = -A_{jj})$:

$$B = \left\{ A_{ij} - A_{jj}; \; i,j = 1,2,\ldots,n \right\} . \qquad (14)$$

It should be mentioned that—obviously—very few special cases like the multi-dimensional Schlögl model appear more complicated in normal form than in the conventional representation with non-zero diagonal elements.

The new replication rate matrix B is used to assign a *colored* graph to the ODE

$$\frac{dx_k}{dt} = x_k \left(\sum_{j=1}^{n} \left\{ B_{kj} x_j - \sum_{\ell=1}^{n} B_{\ell j} x_\ell x_j \right\} \right) ; \quad k = 1,2,\ldots,n . \qquad (15)$$

In the graph the non-zero catalytic terms are shown. Every positive entry of matrix B is represented by a black arrow ($\longrightarrow$), every negative entry by a white arrow ($\Longrightarrow$). The direction of the arrow, $k \leftarrow j$, symbolizes the direction of catalytic action. In order to give examples we show the graphs corresponding to the three matrices

$$\begin{pmatrix} 0 & 0 & 2 & 0 \\ 0 & 0 & 0 & -4 \\ 0 & 1 & 0 & 0 \\ 3 & 0 & 2 & 0 \end{pmatrix}_A \qquad \begin{pmatrix} 0 & 0 & 4 \\ 2 & 0 & 0 \\ 0 & 3 & 0 \end{pmatrix}_B \qquad \begin{pmatrix} 0 & 1 & 1 & 1 \\ 1 & 0 & 1 & 1 \\ 1 & 1 & 0 & 1 \\ 1 & 1 & 1 & 0 \end{pmatrix}_C$$

in Figure 2. The first example (**A**) combines positive and negative couplings in one graph. The second graph (**B**) represents a three-membered hypercycle. Note that hypercycles are easily recognized in the graph representation: they consist of one closed path with exclusively positive coupling terms. In the third example (**C**) all off-diagonal elements have positive entries and the corresponding graph contains all twelve possible arrows.

Intuitively we interpret black arrows in the graphs as activation and white arrows as inhibitory (catalytic) action. One has to keep in mind, however, that the elements of matrix B are differences in rate constants, whereas the conventional interpretation of the signs of catalytic action refers to absolute values.

It is straightforward to try a classification of the dynamics based on the properties of the graphs. Several attempts in this direction are reported in the literature[1,2,20] and a comprehensive review is found in a monograph.[20] Graphs which contain only black arrows $(B_{ij} \geq 0)$ are much easier to study—the corresponding dynamical systems were called *non-hyperbolic* in a previous paper.[17]

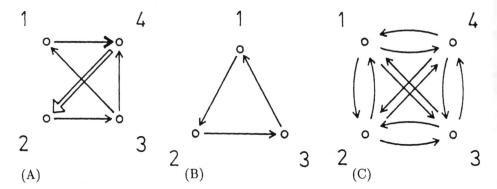

FIGURE 2   Symbolic representation of second-order autocatalytic networks by colored graphs. Black arrows (↔) represent positive entries, white arrows (⇔) negative entries of the coefficient matrix B. The three graphs correspond to the three examples (A, B and C) shown in the text.

Two notions from the theory of directed graphs turned out to be useful: a graph is called *irreducible* if every vertex can be reached from every vertex along a directed path consisting in a sequence of arrows. A *Hamiltonian arc* is a closed path which visits every vertex of the graph exactly once. A catalytic network is called Hamiltonian if it contains an Hamiltonian arc. *Hypercycles* are the simplest Hamiltonian networks since they contain nothing but one Hamiltonian arc. The graph **B** in Figure 2 may serve as an example.

Several useful theorems about permanence of non-hyperbolic catalytic reaction networks ($B_{ij} \geq 0$) were derived. A review of the various proofs is found in the monograph by Hofbauer and Sigmund.[20]

- If a catalytic network with $B_{ij} \geq 0$ is *permanent*, then its graph is *irreducible*.
- *Hypercycles* are *permanent* and thus represent the simplest class of permanent catalytic networks with $B_{ij} \geq 0$ for every given number of vertices (n).

Simplicity refers to the number of arrows in the graph. If we omit or redirect an arrow in the hypercycle, the graph ceases to be irreducible.

- A catalytic network with n=3 and $B_{ij} \geq 0$ is *permanent* if and only if there exists a unique interior fixed point $\hat{x}$ in S_3. At the same time det $B > 0$ holds.
- A catalytic network with $n = 4$ and $B_{ij} \geq 0$ is *permanent* if and only if there exists an interior fixed point $\hat{x}$ in S_4 *and* det $B < 0$ holds.
- If $n \leq 5$ and $B_{ij} \geq 0$, then the graph of a *permanent* catalytic network is Hamiltonian.

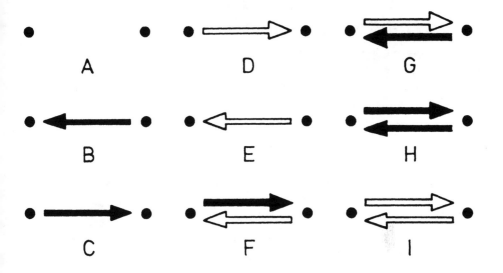

FIGURE 3 Nine different graphs corresponding to all possible cases of replicator equations with $n = 2$. Only the graphs **F**, **G**, **H** and **I** are robust in the sense that they have no zero off-diagonal elements in matrix B and hence do not change on small variations in parameters. Note that the graphs B and C, D and E as well as F and G are equivalent. In essence we are dealing with five different systems.

The intuitive suggestion that all permanent catalytic networks must be Hamiltonian turns out to be wrong in higher dimensions ($n \geq 6$). To give an example the replicator Eq. (11) with

$$B = \begin{pmatrix} 0 & 1 & 0 & 0 & 0 & 3 \\ 2 & 0 & 0 & 0 & 0 & 0 \\ 1 & 0 & 0 & 2 & 0 & 0 \\ 0 & 3 & 1 & 0 & 0 & 0 \\ 0 & 0 & 0 & 3 & 0 & 1 \\ 0 & 0 & 0 & 0 & 1 & 0 \end{pmatrix}$$

is permanent although its graph does not contain a Hamiltonian arc. (Drawing the graph is left as an exercise to the reader!)

SMALL CATALYTIC NETWORKS

For small catalytic networks ($n \leq 3$), full classification of the dynamics is possible. The number of graphs, however, increases drastically with increasing n and this makes complete qualitative analysis intractable for higher-dimensional cases.

THE CASE OF TWO REPLICATORS

For $n = 2$ we have three possible assignments $(+, 0, -)$ of the two relevant matrix elements (B_{12}, B_{21}) which yields $3^2 = 9$ different colored graphs. The graphs determine the phase portraits of the corresponding replicator Eq. (11) uniquely.

All nine graphs are shown in Figure 3. The notion of robustness is very useful in this context: we shall call a graph robust if it is not changed by a small variation in the entries of matrix B which means that no matrix elements of value zero are admitted. Then only $2^2 = 4$ graphs (**F**, **G**, **H** and **I** in Figure 3) remain.

Making use of inherent symmetries in the dynamical systems—for example permutation of vertices—the number of different graphs is reduced to six. Since S_2 sustains only very limited forms of dynamics, the number of distinguishable phase portraits is only four (Figure 4):

1. a flow from the vertices towards a stable fixed point in the interior of S_2,
2. a flow from an unstable fixed point in the interior of S_2 to the vertices,
3. a flow from an unstable vertex to the stable vertex, and
4. the indifferent case in which S_2 is a one-dimensional manifold of fixed points.

The classification can be simplified further by grouping those systems together which are transformed into each other by time-reversal. This implies that cases (1) and (2) fall into the same class and we have only three qualitatively different cases for $n = 2$.

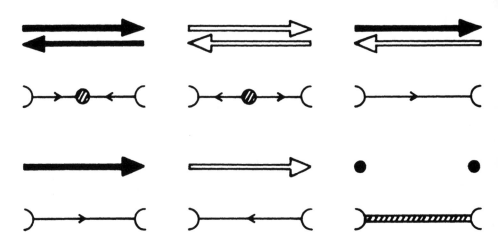

FIGURE 4 Six graphs remaining after elimination of equivalent cases in Figure 3 and the flows on S_2. We distinguish four qualitatively different phase portraits. Stable or unstable fixed point in the interior, flow from one vertex to the other and one-dimensional invariant manifold.

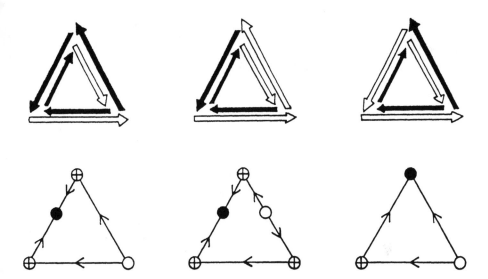

FIGURE 5   The three robust graphs which determine the phase portrait of the replicator equation with $n = 3$ uniquely. The following three symbols are used for fixed points: $\bigcirc$ ... source, $\oplus$ ... saddle and $\bullet$ ... sink.

THE CASE OF THREE REPLICATORS

The system with $n = 3$ comprises already a great variety of dynamical systems classified by $3^6 = 729$ different graphs. These are reduced to 74 essentially different graphs by taking into account the C_3 symmetry operations and time-reversal. In systems with $n = 3$ we encounter a new feature for classification. The graph may or may not determine the phase portrait of the corresponding replicator Eq. (11) uniquely. Here we find 52 graphs with unique phase portraits and 22 graphs which lead to dynamical systems corresponding to two or more phase portraits. Accordingly full characterization becomes already a rather tedious task.[34] As an example for the $n = 3$ case, we show here the three robust graphs which determine the phase portraits uniquely (Figure 5).

Replicator equations with $n = 3$ do not sustain stable closed orbits.[19] But sets of marginally stable oscillations similar to centers in linear ODE's may occur like they do in the conventional Lotka-Volterra equation for two species.

A CHAOTIC ATTRACTOR IN THE SYSTEM WITH FOUR REPLICATORS

The replicator equation with $n = 4$ shows such a rich dynamics that no complete classification could be given as yet. Instead of reviewing various attempts to describe in part this enormous variety, we present an example of deterministic chaos which was recently studied in some detail.[3]

The strange attractor was embedded in a two-dimensional parameter manifold (μ, ν),

$$B(\mu, \nu) = \begin{pmatrix} 0 & 0.5 - 0.437\nu & -0.1 + 0.1\nu & 0.1 + 0.337\nu \\ 1.1 - 0.563\nu & 0 & -0.6 + 0.564\nu & -0.001\nu \\ -0.5 - 0.035\nu & 1 - 0.62\nu & 0 & 0.655\nu \\ 1.7 + \mu - 1.164\nu & -1 - \mu + 0.968\nu & -0.2 + 0.196\nu & 0 \end{pmatrix}$$

and the dynamics was studied as a function of the two parameters. One connected chaotic regime was observed.

Previous studies on Lotka-Volterra equations[3,38] of dimension $n = 3$ reported two distinct strange attractors. The studies of the equivalent replicator equation with $n = 4$ showed that both attractors are just different cross sections through the same chaotic regime: the attractor reported by Vance[38] is equivalent to a strange attractor at $(\mu = 0, \nu = 1)$ in the replicator equation and the one reported by Arneodo, Coullet and Tresser[3] corresponds to the point $\mu = 0$ on the one-dimensional manifold $(-0.11 \leq \mu \leq 0.16, \nu = 0)$. Phase portraits of both attractors are shown in Figures 6 and 7.

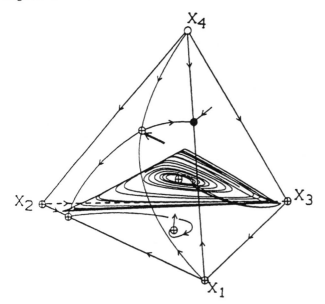

FIGURE 6 Phase portrait and chaotic trajectories of the replicator equation $(\mu = 0, \nu = 1)$ corresponding to the Vance model.[38] Symbols used for fixed points: $\bigcirc$ ... source, $\oplus$ ... saddle and $\bullet$ ... sink.

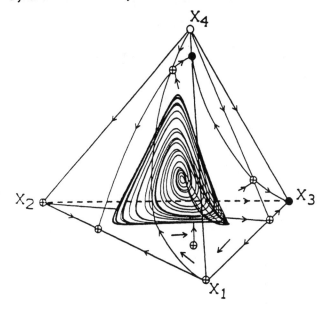

FIGURE 7 Phase portrait and chaotic trajectories of the replicator equation $(\mu, \nu = 0)$ corresponding to the Arneodo-Coullet-Tresser model.[3] Symbols used for fixed points: $\bigcirc$ ... source, $\oplus$ ... saddle and $\bullet$ ... sink.

As an illustrative example of the dynamics in and around the chaotic regime, we consider Poincaré cross sections of the attractors in the (x, y)-plane through S_4 (see Figure 8). The route into chaos along the straight line $\nu = 0$ in the two-dimensional parameter subspace (μ, ν) is shown in Figure 9. In the range of small μ-values $(-0.2 \leq \mu \leq -0.11)$ this route resembles the well-known Feigenbaum sequence very closely. Within the first part of the chaotic regime, we observe the same internal structure—appearance of periodic windows, etc.—as in the discrete logistic equation, $x_{t+1} = kx_t(1 - x_t)$. In contrast to the logistic equation, we observe fully developed chaos nowhere in parameter space. There is always a finite area surrounding the central fixed point $\hat{x}_C$ which is never visited by the trajectories on the attractor. In addition we observe two internal crises at $\mu = 0.054$ and $\mu = 0.168$.

Systematic searches in high-dimensional parameter spaces are extremely demanding on resources of numerical computation. No wonder that questions like how the two-dimensional chaotic regime is continued in the other ten dimensions of the twelve-dimensional parameter space or whether there exist other strange attractors which populate disconnected chaotic regimes cannot be answered as yet.

MUTATION IN NONLINEAR CATALYTIC NETWORKS

Qualitative analysis of the differential Eq. (10) which models catalytic replication and mutation is exceedingly demanding and has been carried out so far only in low-dimensional systems ($n \leq 4$) and even then special assumptions on the matrices of rate constants (A) and mutation frequencies (Q) were inevitable.[35]

With approximations, however, it is possible to achieve a higher degree of generality and to overcome—in principle—the limitation to low-dimensional networks. For this goal a version of perturbation theory[36,37] which can be applied to replicator equations was developed. The perturbation approach becomes exact in the limit of vanishingly small mutation rates. The kinetic equations for replication and mutation (10) are rewritten in such a way that error-free production and mutant formation appear as additive contributions:

$$\frac{dx_k}{dt} = \mathcal{R}_k(\mathbf{x}) + \mathcal{M}_k(\mathbf{x}, \varepsilon) ; \quad k = 1, 2, \ldots, n \tag{15}$$

with

$$\mathcal{R}_k(\mathbf{x}) = x_k \left(\sum_{j=1}^{n} \left\{ A_{kj} x_j - \sum_{\ell=1}^{n} A_{\ell j} x_\ell x_j \right\} \right)$$

and

$$\mathcal{M}_k(\mathbf{x}, \varepsilon) = \sum_{\ell=1}^{n} \sum_{j=1}^{n} \left(Q_{k\ell}(\varepsilon) A_{\ell j} x_\ell x_j - Q_{\ell k}(\varepsilon) A_{kj} x_k x_j \right) .$$

The first part—the unperturbed replicator field—is identical to Eq. (11). The mutation field $\mathcal{M}$ is considered as perturbation with ε being a perturbation parameter which is still to be defined. As mentioned before the replicator part is insensitive to the addition of constants to the reaction rate parameters[17]: $A'_{ij} \rightarrow A_{ij} + \Delta$. In order to make the perturbation approach applicable to the replicator equation, a sufficiently large value of the additive constant Δ has to be chosen such that all entries of matrix A' are positive.

In order to be able to apply perturbation theory, we describe the matrix elements of the mutation matrix Q in a Taylor series:

$$Q_{ij}(\lambda) = \sum_{m=0}^{\infty} \frac{1}{m!} \frac{\partial^m Q_{ij}(0)}{\partial \lambda^m} \lambda^m = \sum_{m=0}^{\infty} Q_{ij}^{(m)} \lambda^m .$$

Introduction of a *mean mutation rate* ε

$$\varepsilon = \lambda \cdot \frac{1}{n(n-1)} \sum_{i=1}^{n-1} \sum_{j>i}^{n} Q_{ij}^{(1)}$$

and a kind of *normalized* expansion series of the mutation matrix

$$\Omega_{ij}^{(m)} = \left(\frac{\lambda}{\varepsilon} \right)^m Q_{ij}^{(m)}$$

allows to write the mutation field in the following form:

$$\mathcal{M}_k(\mathbf{x}, \varepsilon) \;=\; \sum_{m=1}^{\infty}\left(\sum_{\ell=1}^{n}\sum_{j=1}^{n}\Big\{\Omega_{k\ell}^{(m)} A_{\ell j} x_{\ell} x_j - \Omega_{\ell k}^{(m)} A_{kj} x_k x_j\Big\}\cdot \varepsilon^m\right).$$

This expansion is now in appropriate form for perturbation analysis.

The influence of mutation on the position of the kth fixed point of the unperturbed replicator Eq. (11) $(\widehat{\mathbf{x}}_k^{(0)})$ is expressed by

$$\widehat{\mathbf{x}}_k(\varepsilon) \;=\; \widehat{\mathbf{x}}_k^{(0)} + \varepsilon\cdot\mathbf{d_k} + O(\varepsilon^2).$$

As we have shown recently[35,36] the shift vectors $\mathbf{d_k}$ can be obtained from

$$J(\widehat{\mathbf{x}}_k^{(0)})\cdot\mathbf{d_k} \;=\; -\,\mathcal{M}_k(\widehat{\mathbf{x}}_k^{(0)}, 0),$$

where $J(\mathbf{x})$ is the Jacobian of the unperturbed system: $J \doteq \{J_{ij} = \partial\mathcal{R}_i/\partial x_j\}$.

Two theorems can be derived from Eqs. (20) and (21) which allow to predict changes caused by infinitesimal mutation rates ε in the positions of the fixed points of replicator equations which lie on the boundary of the simplex S_n and whose Jacobian has only non-zero eigenvalues.

- An asymptotically stable fixed point migrates from the boundary of the simplex S_n into the interior.
- An unstable fixed point migrates from the boundary of the simplex S_n outwards and thus leaves the physically accessible part of $\mathbf{R}^n$.

Changes in phase portraits can be predicted by means of these two theorems provided the unperturbed replicator system is known.

The multi-dimensional Schlögl model provides a useful example to study the influence of mutation. All corners of the simplex S_n are stable fixed points in the error-free replicator equation and hence move into the interior with increasing mutation rates ε. All the other fixed points which lie on the boundary of S_n—one on each edge, each face, etc.—are unstable in the replicator field $\mathcal{R}$ and hence move outside the physically meaningful domain. At small mutation rates every replicator is surrounded by a mutant cloud which in the language of reaction dynamics means that the n stable stationary states lie close to the corners of S_n. At higher mutation rates the multi-stable system becomes simpler since stable fixed points vanish—one after the other or simultaneously depending on details of the matrices A and Q—until finally only one stable fixed point inside S_n remains. The critical mutation rates at which the individual fixed points disappear are related to the error-threshold of the quasi-linear replication network.

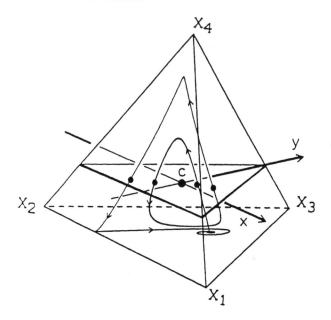

FIGURE 8   Orientation of the attractor relative to the (x, y)-plane which is chosen for the Poincaré cross section. The fixed point in the center of S_4 is denoted by $\hat{\mathbf{x}}_C = (1/4, 1/4, 1/4, 1/4)$.

As a second example we consider the strange attractor discussed in the previous section. Figure 7 shows two asymptotically stable fixed points, $\hat{\mathbf{x}}_3$ and $\hat{\mathbf{x}}_{14}$. All other fixed points at the boundary of S_4 are unstable. Hence only the first two points move into the interior and become stable stationary states of the system. In Table 1 we present the shift vectors $\mathbf{d_k}$ of some fixed points of the chaotic model replicator equation at $\nu = 0$ as they were obtained from Eq. (21) Shifts inward and outward from the boundary of S_4 are easily verified for the chaotic attractor at $\mu = 0$.

In order to study the chaotic attractor also at larger error rates, we explored the parameter and mutation space (μ, ε) by integration of the corresponding differential equation. Figure 10 summarizes the changes in the chaotic attractor due to mutation. The main result of this study is that—at least this particular type of—chaotic dynamics is very sensitive to mutation: mutation probabilities as small as 2×10^{-5} are sufficient to destroy the strange attractor. At higher mutation rates—ε larger than some 10^{-3}—the interior fixed point is stable. Proceeding in opposite direction we find that a decrease in the mutation rate leads to a sequence of period doublings which eventually ends up in chaos. A highly complex bifurcation structure arises on the introduction of mutation near the limit crisis of the unperturbed attractor. At the present stage of resolution we cannot say whether the bifurcation set forms a fractal or not.

It is not surprising that even small mutation rates destroy the chaotic attractor since the mutation term causes the saddle focus $\hat{\mathbf{x}}_{123}$ to move into the unphysical range outside of the simplex S_4.[35] Therefore the trajectories can no longer come sufficiently close to the saddle and Šilnikov's mechanism breaks down (see Figures 7 and 8). Furthermore the fixed point $\hat{\mathbf{x}}_{14}$ moves into the interior of S_4 and thus the limit crisis occurs earlier.

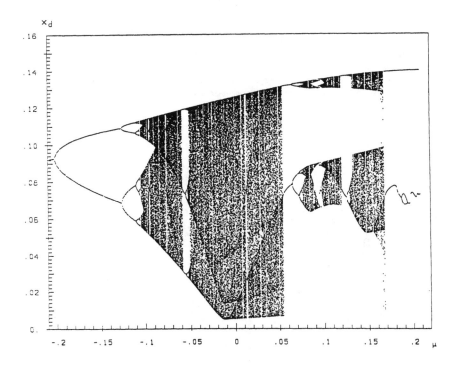

FIGURE 9 A Poincaré map of the attractors with $\nu = 0$ as a function of the parameter μ. The map is projected on the x-axis defined in Figure 8. The origin $(x = 0)$ coincides with the fixed point in the center of S_4 $(\widehat{\mathbf{x}}_C)$.

Direct qualitative analysis of catalytic networks with mutation was carried out for special cases of low-dimensional systems also based on different model assumptions. As an example we mention a study by García-Tejedor, Moran and Montero on the hypercycle equation.[16] In this approach a mutant cloud was assumed to be formed around each element of the catalytic network. In contrast to the model applied here mutants are not considered to be members of the replicating ensemble.

TABLE 1 Examples of shift vectors for some fixed points caused by mutation in the replicator model with $\nu = 0$. All mutation rates were assumed to be equal $(Q_{ih} = \epsilon \forall i \neq j)$ and the results were obtained by perturbation theory.[35,36]

k^1	Fixed point $\widehat{\mathbf{x}}_k^{(0)}$	Shift vector $\mathbf{d_k}$
1	$\begin{pmatrix} 1 \\ 0 \\ 0 \\ 0 \end{pmatrix}$	$\frac{16}{11(10\mu+17)} \cdot \begin{pmatrix} -(60\mu + 47) \\ -5(10\mu + 17) \\ 11(10\mu + 17) \\ -55 \end{pmatrix}$
3	$\begin{pmatrix} 0 \\ 0 \\ 1 \\ 0 \end{pmatrix}$	$\frac{40}{3} \cdot \begin{pmatrix} 6 \\ 1 \\ -10 \\ 3 \end{pmatrix}$
14	$\frac{1}{2(5\mu+9)} \begin{pmatrix} 1 \\ 0 \\ 0 \\ 10\mu + 17 \end{pmatrix}$	$\frac{810\mu+1457}{2(5\mu+3)(5\mu+9)(5\mu+11)(10\mu+17)} \cdot$ $\begin{pmatrix} (5\mu + 11)(100\mu^2 + 265\mu + 167) \\ (5\mu + 9)(5\mu + 11)(10\mu + 17) \\ (5\mu + 3)(5\mu + 9)(10\mu + 17) \\ -(1000\mu^3 + 4875\mu^2 + 7730\mu + 3979) \end{pmatrix}$
123	$\frac{1}{32} \begin{pmatrix} 10 \\ 7 \\ 15 \\ 0 \end{pmatrix}$	$\frac{43}{112(3\mu+5)} \cdot \begin{pmatrix} -6(9\mu - 13) \\ 87\mu + 61 \\ -33\mu + 533 \\ -672 \end{pmatrix}$

[1] The individual fixed points are denoted by indicating the nonvanishing coordinates as subscript. As can be verified easily the fixed points $\widehat{\mathbf{x}}_3$ and $\widehat{\mathbf{x}}_{14}$ migrate into the interior of S_4, $\widehat{\mathbf{x}}_1$ and $\widehat{\mathbf{x}}_{123}$ move outwards.

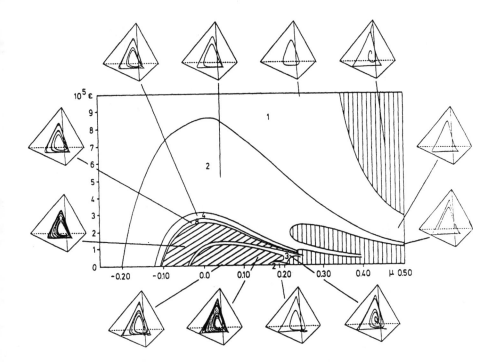

FIGURE 10 Extension of the chaotic regime into the range of small mutation rates.
The dynamics of the replicator model ($\nu = 0$) in the range $-0.25 < \mu < 0.5$ and
$0 \leq \varepsilon < 10^{-4}$ is shown. Typical shapes of the attractors are sketched. In the vertically
hatched range trajectories to a stable point on the edge $\overline{14}$. Periodic attractors are
characterized by period multiplicities ($1 \equiv$ simple limit cycle). Chaotic dynamics is
observed in the zone with slanted hatching. Note that a Feigenbaum-sequence-like path
with successive period doublings ($1 \rightarrow 2 \rightarrow 4 \rightarrow 8 \rightarrow \dots$) is observed in the approach
towards chaos from high to low mutation rates ε.

CONCLUDING REMARKS

In this contribution we have not dealt explicitly with degradation of replicators into
low energy materials **B** as described by the reaction

$$\mathbf{I}_k \xrightarrow{D_k} (\mathbf{B}) ; \quad k = 1, 2, \dots, n .$$

Under the conditions of *in vitro* RNA replication experiments this is indeed a
realistic assumption since there hydrolytic degradation is a very slow process. The
neglect may be less well justified in other systems but then it is equivalent to

the assumption of identical rate constants $D_1 = D_2 = \ldots = D_n = D$ because any unspecific degradation term $x_k \cdot D$ may be absorbed into the flux $x_k \cdot \Phi$.

So far we have not considered spatio-temporal phenomena here. This was done for two different reasons: firstly, their analysis requires the study of reaction-diffusion equations—a special class of parabolic PDE's—which are much less well understood than ODE's and secondly, only very few studies on the class of processes we mainly are interested here have been carried out. We restrict this discussion to a very brief comment therefore.

Quasi-linear replicator networks do not form spatial patterns under homogeneous Neumann or no-flux boundary conditions. Non-stationary spatial phenomena are nevertheless possible. As an example we mention travelling waves which were studied in theory and observed in a suitable experimental setup.[4]

Essentially nonlinear catalytic networks should be able to exhibit the entire spectrum of spatio-temporal phenomena comprising formation of travelling waves, stationary spatial patterns, and oscillating patterns. Numerical integrations support this suggestion.

In the study of replicator and Lotka-Volterra systems the problem of permanence played an predominant role. In prebiotic chemistry, coexistence of replicating molecules as a result of suppressed selection has been studied extensively.[10] Other applications are the study of various forms of symbiosis in biology and the question of coexistence of species in the predator-prey models of theoretical ecology.

Mutation is an *unavoidable* by-product of replication. In contrast to the well-studied quasi-linear replication mutation networks very little was known about nonlinear catalytic networks with mutation. Quantitative answers to the questions if and how complex replication dynamics is changing under the influence of mutation is of obvious importance for biological applications. The two theorems reported here are a first step towards a general understanding of these systems. The observed simplification of replication dynamics as a consequence of increased mutation rates can be interpreted by a straightforward argument: the effect of nonlinearities is often visualized by the introduction of delays into essentially linear systems. Delays may lead to complex dynamics, in particular to the formation of temporal or spatial patterns. Mutations represent parallel pathways producing other species directly, i.e., along *short-cuts*. Reactions proceeding along such short-cuts of the catalytic networks compensate for delays. Clearly such an effect becomes stronger with increasing mutation rate.

ACKNOWLEDGMENTS

The work reported here was supported financially by the *Fonds zur Förderung der wissenschaftlichen Forschung in Österreich* (Projects No. 5286 and No. 6864), the *Stiftung Volkswagenwerk* (B.R.D.), and the *Hochschuljubiläumsstiftung Wien*.

Numerical computations were performed on the IBM 3090 mainframe of the *EDV-Zentrum der Universität Wien* as part of the IBM Academic Supercomputing Program for Europe (EASI). Assistance in preparation of figures by Mr. J. König and Drs. Wolfgang Schnabl, Christian Forst, and Peter Stadler is gratefully acknowledged.

REFERENCES

1. Amann, E., and J. Hofbauer. "Permanence in Lotka-Volterra and Replicator Equations." In *Lotka-Volterra-Approach to Cooperation and Competition in Dynamic Systems*, vol. 23, edited by W. Ebeling and M. Peschel. Math. Research, Berlin: Akademie-Verlag, 1985, 23–34.
2. Amann, E., and J. Hofbauer. "Permanence in Population Dynamics." In *Dynamical Systems and Environmental Models*, edited by H. G. Bothe, W. Ebeling, A. B. Kurzhanski, and M. Peschel. Berlin: Akademie-Verlag, 1987, 58–66.
3. Arneodo, A., P. Coullet, and C. Tresser. *Phys. Lett.* **79A** (1980):259–263.
4. Bauer, G. J., J. S. McCaskill, and H. Otten. *Proc. Natl. Acad. Sci. USA* **86** (1989):7937–7941.
5. Dawkins, R. *The Selfish Gene.* Oxford: Oxford University Press, 1976, 13–21.
6. Demetrius, L, P. Schuster, and K. Sigmund. *Bull. Math. Biol.* **47** (1985):239–262.
7. Eigen, M. *Naturwissenschaften* **58** (1971):465–523.
8. Eigen, M., and P. Schuster. *Naturwissenschaften* **65** (1978):7–41.
9. Eigen, M., and P. Schuster. *The Hypercyle—A Principal of Natural Self-Organization.* Berlin: Springer Verlag, 1979.
10. Eigen, M., and P. Schuster. *J. Mol. Evol.* **19** (1982):47–61.
11. Eigen, M., J. McCaskill, and P. Schuster. *J. Phys. Chem.* **92** (1988):6881–6891.
12. Eigen, M., J. McCaskill, and P. Schuster. *Adv. Chem. Phys.* **75** (1989):149–263.
13. Ewens, W. J. *Mathematical Population Genetics.* Berlin: Springer Verlag, 1979.
14. Fontana, W., and P. Schuster. *Biophys. Chem.* **26** (1987):123–147.
15. Fontana, W., W. Schnabl, and P. Schuster. *Phys. Rev. A* **40** (1989):3301–3321.
16. Garcia-Tejedor, A., F. Moran, and F. Montero. *J. Theor. Biol.* **127** (1987):393–402.
17. Hofbauer, J., P. Schuster, K. Sigmund, and R. Wolff. *SIAM J. Appl. Math.* **38** (1980):282–304.
18. Hofbauer, J., P. Schuster, and K. Sigmund. *J. Math. Biol.* **11** (1981):155–168.
19. Hofbauer, J. *Nonlinear Analysis* **5** (1981):1003–1007.

20. Hofbauer, J., and K. Sigmund. *The Theory of Evolution and Dynamical Systems*. London Mathematical Society, Student Texts, vol. 7. Cambridge (U.K.): Cambridge University Press, 1988.
21. McCaskill, J. *Biol. Cybernet.* **50** (1984):63–73.
22. Nowak, M., and P. Schuster. *J. Theor. Biol.* **137** (1989):375–395.
23. Phillipson, P. E., and P. Schuster. *J. Chem. Phys.* **79** (1983):3807–3818.
24. Phillipson, P. E., P. Schuster, and F. Kemler. *Bull. Math. Biol.* **46** (1984):339–355.
25. Shahshahani, F. "A New Mathematical Framework for the Study of Linkage and Selection." *Memoirs Amer. Math. Soc.*, vol. 211. Providence, RI: AMS, 1979.
26. Schlögl, F. *Z. Phys.* **253** (1972):147–161.
27. Schnabl, W., P. F. Stadler, C. Forst, and P. Schuster. Submitted to *Physica D*, 1990.
28. Schuster, P., and K. Sigmund. *J. Theor. Biol.* **100** (1983):533–538.
29. Schuster, P., and K. Sigmund. *Ber. Bunsenges. Phys. Chem.* **89** (1985):668–682.
30. Schuster, P. *Physica* **22D** (1986):100–119.
31. Schuster, P. *Physica Scripta* **35** (1987):402–416.
32. Schuster, P., and K. Sigmund. *Math. Biosci.* **95** (1989):37-51.
33. Schuster, P. This volume.
34. Stadler, P. F., and P. Schuster. *Bull. Math. Biol.* **51** 1990.
35. Stadler, P. F., W. Schnabl, C. Forst, and P. Schuster. Submitted to *Math. Biosc.*, 1990.
36. Stadler, P. F. *Selection, Mutation and Catalysis*. Ph.D. Thesis. Universität Wien, 1990.
37. Stadler, P. F., and P. Schuster. Submitted to *J. Math. Biol.*, 1990.
38. Vance, R. R. *Amer. Natur.* **112** (1978):797–813.
39. Vrba, E. S. "Levels of Selection and Sorting with Special Reference to the Species Level." In *Oxford Surveys in Evolutionary Biology*, vol.6, edited by P. H.Harvey and L. Partridge. Oxford: Oxford University Press, 1989, 114–115.

Index

Printed in the United States
by Baker & Taylor Publisher Services